Research Methodology in Social Sciences

NIPA® GENX ELECTRONIC RESOURCES & SOLUTIONS P. LTD.
New Delhi-110 034

Research Methodology in Social Sciences

Shridhar Patil

and

Aditya

NIPA® GENX ELECTRONIC RESOURCES & SOLUTIONS P. LTD.

New Delhi-110 034

NIPA® GENX ELECTRONIC
RESOURCES & SOLUTIONS P. LTD.

101,103, Vikas Surya Plaza, CU Block
L.S.C. Market, Pitam Pura, New Delhi-110 034
Ph : +91 11 27341616, 27341717, 27341718
E-mail: newindiapublishingagency@gmail.com
Web: www.nipabooks.com

For customer assistance, please contact
Phone: + 91-11-27 34 17 17
Fax: + 91-11- 27 34 16 16
E-Mail: feedbacks@nipabooks.com

ISBN: 978-81-19254-84-2

Composed and Designed by NIPA®.

Preface

This foundation manual was developed for the use of graduate and post graduate research scholars of the Agricultural Extension/ Extension Education/ Rural Development. In this book we have made an attempt to clearly illustrate the fundamental concepts related to the aspect of social research in the context of Extension Education. In order to ease the understanding of basic concepts, widely used simplest research methods and tools and techniques of data collection, we have divided the book into four parts, Foundations of social research, Research methods, Tools and techniques of data collection and Data processing and report writing.

Foundations of social research deals with universal and basic units of social research like scientific approach, meaning, process and development of scientific research problem. It also deals with defining and measurement of variables and testing of reliability and validity of measuring instruments. In this manual attempts are made to illustrate these concepts with the relevant examples.

Research Methods section deals with the three major research methods used in extension education/ Agricultural extension, namely Survey research, Action research and case study. This section discusses in detail the process, relative advantages and limitations of each of these three methods. There are numerous research methods used in social research. Rather than discussing all these research methods in brief, we made an attempt to discuss in detail the most widely used research methods in the field of agricultural extension.

Tools and techniques of data collection deals with situation suitability, relative advantages and limitation of various data collections techniques like face to face interview, mailed questionnaire, observation method, content analysis, sociometry and projective methods.

Data processing and report writing section deals with making the collected data amenable for statistical analysis i.e. coding. This section discusses in detail the various types of codes and their utility. It also deals with formulation and testing of hypothesis and writing of the research report. Since this manual is expected to be used by post graduate and doctoral research scholars of agricultural extension/ extension education, the scope of research report is limited to Thesis writing and the other types of research reports like technical report, working paper etc. are not discussed in this manual.

We hope that, this book will be instrumental in developing the understanding of scientific research and its process among the students of agricultural extension/ extension education.

Authors

Contents

PART I: FOUNDATIONS OF SOCIAL RESEARCH

Chapter 1

Concept of Scientific Research and Scientific Approach

Term '*Scientific research*' is too broad to be explained by using one or two definitions. One should go through the different expert views about scientific research in order to internalize the concept of scientific research. In essence a scientific research is characterized by its rigor, testability, replicability, precision, confidence, objectivity, generalizability and parsimony.

Research: Uma Sekaran (2006) defined research as an organized, systematic, data-based, critical, objective, scientific inquiry or investigation into a specific problem, undertaken with the purpose of finding answers or solutions to it.

Scientific Research: Scientific research is systematic, controlled, empirical and critical investigation of hypothetical preposition about the presumed relations among natural phenomenon (Kerlinger, 1973).

Scientific research is characterized by the set of features in its planning, execution, observation and interpretation. These features listed below, differentiate scientific investigation from other forms of investigation.

How scientific research differs from other forms of investigation?

i. *Based on observable and measurable facts*: This means that the scientific research and interpretation of the results of the scientific research are based on the facts that can be observed and measured either directly or indirectly. The use of metaphysical explanations to interpret the cause-effect relationships is not acceptable in scientific research.

ii. *Observations are highly disciplined:* This means that there are well established clear guidelines for what to observe, how to observe, how to measure and how to record the observations.

iii. *Free from subjective bias*: This refers to the property of scientific research wherein the process of the research and interpretation of the research results are drawn on the basis of objective criteria and are free from subjective biases of the researcher, sponsoring agency or any other external influence.

iv. *Subjected to open court of empirical enquiry*: Results observed and interpretations made by scientific research are based on empirical facts and are open for evaluation, testing, criticism and refinement by anybody. Others are free to repeat the scientific research and compare the results with that of previous research and if any discrepancies are found, then science seeks for explanation and self correction.

v. *Systematic and controlled process*: Scientific research is carried out in controlled conditions so as to minimize the effect of extraneous variables and increase the precision in the detection of cause effect relationship.

Scientific approach: Based on the John Dewey's analysis of reflective thinking, we can define the scientific approach as a step by step and systematic method of identifying & defining a problem, formulating tentative hypothetical relation, reasoning and deducing hypothesis to testable concrete form and then subjecting it to experimentation, testing & observation so as to draw valid and objective conclusions.

Steps involved in scientific approach (Dewey, 1998)

1. *Problem-obstacle idea*: This is the initial stage of scientific approach. In this stage an individual perceives a problem but faces obstacle in explaining and defining it. He refines problem and comes out with well-defined and manageable scientific problem. *Scientific problem* is interrogative statement that asks what relationship exists between two or more natural phenomenon.

2. *Hypothesis*: After defining problem, researcher uses his experience, intuition, knowledge and peer group knowledge to formulate a tentative relational statement among two or more natural phenomena. As Kerlinger, (1973) states, *Hypothesis is a conjectural statement, a tentative proposition, about a relation between two or more phenomena or variable*. Alternatively, one can also formulate the testable hypothesis about existence, distribution, intensity and frequency of a natural phenomenon. Hypothesis need not always test the cause-effect relation. It can be formulated even to test just the covariation among two or more natural phenomenon or it may also test whether the actual observations are same as assumptions/predictions.

3. *Reasoning and deduction*: At this stage, the researcher deduces the hypothesis to more basic underlying problem and even the problem itself may be redefined/changed as a result of reasoning and deduction leading to testable, operationalized and basic problem.

4. *Observation-test experimentation*: Here the researcher subjects the deduced implications of hypothesis to experimentation under controlled conditions via manipulation, randomization and local control. Researcher observes the test outcomes objectively and carefully and carries out analysis and infers the results to draw objective conclusions. Note that we test deducted implications of hypothesis and not the hypothesis itself.

Scientific research is a cyclic process, it revaluates the theory by comparing the facts with predictions, seeks explanation for discrepancy between theoretical prediction and factual observation; and facilitates self correction of scientific knowledge.

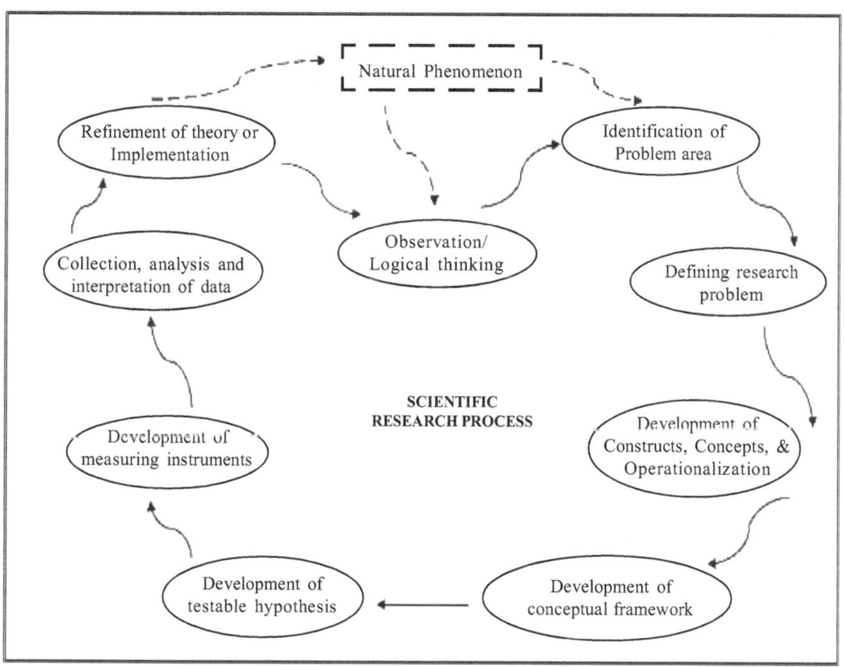

Exercise

a. Prepare a brief check list for evaluating research process.

b. Select any scientific publications (preferably thesis) and evaluate the process of scientific research followed by the researcher using the check list which you already prepared.

References

Kerlinger, F. (1973). *Foundations of behavioural research* (2[nd]ed.). New York: Holt, Rinehart and Winston.

Sekaran, U. (2006). *Research methods for business: A skill-building approach* (4th ed.). New York: Wiley.

Dewey, J. (1998). *How we think: A restatement of the relation of reflective thinking to the educative process.* Boston: Houghton Mifflin.

Chapter 2

Selection and Formulation of Research Problem

Rationally selected and contextually defined research problem is the foundation to the successful research project. If the selected research problem is not in line with the interests of the researcher, available time, resources, infrastructure & knowledge base then achieving success in research process remains highly uncertain. Consequently, in addition to hindering of professional growth of the researcher, it also results in wastage of time, money and human effort. Most importantly, it may sometimes add up incorrect knowledge into scientific base, hindering the process of scientific progress. Hence the identification of the research problem and designing the problem statement is the most important step in the research process.

What is a research problem?

Problem is an identified state of difficulty which needs to be addressed. *Scientific problem* is a well defined condition of lack of information required for scientific description, explanation and understanding of an observable phenomenon or lack of theoretical basis required for prediction of a natural phenomenon.

Sources of research problem

1. *Professional experience*: Own experience or the experience of others help in scanning the scientific environment for identification of new problems and refinement of old ones.
2. *Scientific literature*: Review of scientific literature helps in understanding the current advances in knowledge and reveals the discrepancies between theoretical predictions and facts. Further it delineates the need for knowledge advancement. Thus knowledge gap identified via review of literature is one of the important sources of identifying research problem.

3. *Suggestions from and discussion with experts*: Discussion with and suggestions from experts provides an experience based insight into the problem area and reveals the past research trends, existing knowledge gaps, methodological possibilities and the approaches for identifying the specific topic in area of researcher's personal interest.

4. *Observation*: A researcher can be characterized by his/her ability to observe the facts, interrelate these facts and synthesize them into meaningful pattern and relate them with his existing knowledge base. Once the researcher relates facts with his knowledge, he can also identify the knowledge gap based on which one can design the research problem and problem statement.

Process of Identification and development of research problem

1. *Identify the broad area of personal interest*: Scientific research demands commitment, involvement, honest efforts and creative thinking on the part of researcher. If the researcher is made to involuntarily work on certain research problem, he may not show dedication, honest effort and creative thinking to address the research problem. Hence for the success of the project, researcher's *personal interest* about the problem area is of greater importance. For example an agricultural extension student can consider 'farmers' training as his broad area of interest. The personal interest comes from an individual's need to know more about a specific subject. At this stage problem area is too broad and nonspecific. Personal problem emphasises on question *"What I want to know?"*

2. *Transform broad personal interest into researchable interest*: One should always transform personal interest into *researchable interest*. In this stage the question is transformed into "*what those in this academic field know about the subject and what do they need to know?*" This transformation demands extensive review of literature to understand the current state of knowledge and identify the knowledge gap in the area of interest.

3. *Narrow down the problem area to specific problem*: Too broad topic creates vague and too narrowed down topic is practically meaningless. Hence the topic should be specified and narrowed down optimally. Topic should be selected and narrowed down considering the time, resources, knowledge and technological base available with the researcher. For example, suppose an agricultural extension student is interested in farmers' training, then he/she can choose broad problem area 'How to make extension training programme more effective'. But this is too general

topic to investigate on. This can be narrowed down to '*effectiveness of interactive assistance based training programme in enhancing farmers' adoption of technology*'.

In order to make the observation, measurement and generalization possible, one must define the boundaries of the problem by defining the research territory, experimental units and the sampling frame etc. The narrowed down problem should be rationale, researchable, manageable, have theoretical basis and should facilitate generalization of results.

4. *Develop Problem (need) statement:* Statement of the research problem describes clearly the nature and extent of the problem and reveals the factors contributing to the problem or the circumstances creating the need. Note that, the term 'problem statement' is used for research-oriented proposals and the term 'need statement' is used for proposals that seek funding for programs or services.

Components of problem / Need statement

a. *The nature and extent of the need/problem*: Background of the problem, what is the problem, who is experiencing it, problem severity and retrospective assessment of the problem.

b. *Factors contributing to the problem or conditions*: This explains the supposed causes of the problem; *e.g.* limited resources or access to services, attitude, lack of awareness, skill, knowledge etc.

c. *Consequences of the problem:* This section entails the impact of the problem on the stakeholders, their family, and the community at large. This section also entails what improvements can be made by addressing the problematic situation.

d. *Approaches for addressing the problem*: This section details the strategies can be used to address the problem.

e. *Limitations of the study:* This describes the assumptions of the study, methodological loopholes in the study and the limits of generalization.

Criteria of an ideal scientific problem statement

1. State/question relationship between two or more variables or it should seek for exploring the observable character of the phenomenon or object under investigation

2. The problem should be significant
3. Research problem should be in line with the time and resources available with the researcher
4. It should be stated clearly and unambiguously
5. It shouldn't be too general or too specific
6. Research problem should have implications for empirical testing
7. Scientific problem should be free from value judgment aspect.

Exercise

a. Identify broad research area of your interest, transform broad personal interest area into research interest and then narrow down research topic to specific research problem
b. Write down the problem statement for the research problem that you have developed.

_ _

_ _

_ _

_ _

_ _

_ _

_ _

_ _

_ _

_ _

_ _

_ _

Reference

Coley, S. and Scheinberg, C. (2008). *Proposal Writing: Effective Grantsmanship* (3rd ed.). Thousand Oaks, California: SAGE Publications.

Chapter 3

Conceptual Framework: Meaning and its Development

Whenever we design a scientific study it is designed in a specific contextual background to solve one or another scientific problem. Subsequently it also involves a set of variables/phenomenon operating in a particular context, set of relationships among these variables and the conditions under which they operate. Certainly, to address an unsolved scientific problem we need to carry out research by employing well defined research methodology and have an idea about possible outcomes. For example when a extension scientist is asked to study the impact of a novel programme, he needs to understand the background and the context under which the programme was introduced, what are the developmental inputs given under the programme, to whom, when (variables), with what purpose (context), what kind of impact he/she is supposed to investigate (presumed relationship), what methods he is going to use to investigate the impact of developmental programme (research methodology) and what are the possible outcomes of the investigation and how programme interventions are related to the outcome (relationship).

All these aspects of a scientific research need to be brought together in a single framework, either in a narrative or graphical format, so that researcher can think, refine and act upon it. Conceptual framework facilitates the above said process of bringing together various components of the systematic conceptualization. In this chapter we will discuss the concept of conceptual framework of scientific research in detail–what is the conceptual framework? Why we need it? What are its components? How to develop a conceptual framework?

Meaning of conceptual framework

K. F. Punch (1998) defines conceptual framework as the "*Conceptual status of the things being studied and their relationship to each other*". It is primarily a conception or model of what is out there that you plan to study, why, a tentative theory of the phenomena that you are investigating (Ravitch and Riggan, 2012). According to Miles and Huberman (1994), *It is the framework that explains, either graphically or in narrative form, the main things to be studied—the key factors, concepts, or variables—and the presumed relationships among them.* According to Maxwell (1996) conceptual framework is the system of concepts, assumptions, expectations, beliefs, and theories that supports and informs your research.

Based on these definitions, we can understand the conceptual framework as the *interrelated system of the research problem, the context, variables and phenomenon under the purview of research problem, presumed relationships that exists (tentative theory) between these variables/phenomenon and the research methods we are going to use to investigate these presumed relationships with the support of existing theories and assumptions.*

Why we need conceptual framework?

A well-conceived conceptual framework is influenced by and at the same time influences the research process at all the levels and at all the stages. It is important to remember that, thinking about your conceptual framework and actually building it is an iterative process. It is revised, reflecting emergent findings and new insights. Conceptual framework enables researchers to make reasoned defensible choices, match research questions with those choices, align analytic tools with research questions, and thereby guide data collection, analysis, and interpretation (Ravitch and Riggan, 2012).

Conceptual framework plays central role throughout the entire research process, and, most important, in the final analysis. Without conceptual development and refinement a study could remain weakly conceptualized, under theorized, and less generative of quality data (Bloomberg and Volpe, 2012).

Practical importance of conceptual framework

a. It provides a theoretical clarification on research context, what researcher is intended to investigate.

b. Conceptual framework assists in defining the research problem, as well as in selection of appropriate bodies of literature for review.

c. Developing a conceptual framework compels researchers to be explicit about what they are doing, and also helps them to be clear about the required data, nature of data and data analysis procedures.

d. Conceptual framework becomes the lens through which the research problem is viewed, providing a theoretical overview of intended research as well as some sort of methodological order within that process.

e. Conceptual framework acts as bridge between paradigms that explain the research issue and the actual practice of investigating that issue.

f. It serves as a filter for developing appropriate research questions.

g. Conceptual framework gives coherence to the research activities by providing traceable connections between theoretical perspectives, research strategy and design, fieldwork, and the conceptual significance of the evidence.

h. Conceptual framework makes it easy for researcher to communicate and for scientific readers to understand the purpose, context and logical justification behind the investigation of given phenomenon or presumed relationship among set of phenomenon.

Components of conceptual framework - Blaxter, Hughes, and Tight (1996)

a. Context of the research

b. Variables /phenomenon

c. Set of relationships among these variables

d. Conditions under which they operate

e. Theories and beliefs that support and inform your research

f. Research methods

g. Expectations.

Developing Conceptual framework

I. Sources of input for developing conceptual framework

a. *Experiential knowledge of student and supervisor:* It includes the researchers' technical knowledge, his/her research background and the personal experience and familiarity with the data (particularly for qualitative) from past researches.

b. *Review of literature*: Prior related theories, past research works and the approaches/lines of work used in some other field of research.

c. *Discussion with experts and brainstorming*: This includes the opinion, insight and critical analysis of the researchers' conceptualization of research strategy and the supposed method of investigation.

II. Process of development of conceptual framework

a. Explore the phenomena/event of your interest in the scientific perspective.

b. Identify the key variables/factors/concepts which appear in the review of literature on the subject of interest.

c. Explore the literature to understand the current level of scientific understanding about the phenomenon of interest and identify the missing links and knowledge gap.

d. Develop a scientific problem based on the knowledge gap.

e. Specify the context and territory of your research problem statement.

f. Identify the potential constructs/phenomenon/factors that might be the possible cause/effect of the phenomenon/event under investigation.

g. Specify the presumed relationships (expectation) between the identified variables/phenomenon.

h. Specify the theories and beliefs that support your argument (expectation).

These steps are not always sequential. You will find yourself going back and forth among them throughout the research and writing process. It is important to remember that, thinking about your conceptual framework and actually building it is an iterative process. As such, an initial conceptual framework can—and most likely will—be revised, reflecting emergent findings and new insights.

Checklist for reviewing the developed conceptual framework

Sl. No.	Check	Remark
1.	Does the conceptual framework draw on theory, experience, and research?	
2.	Does it demarcate the holistic territory of the research?	
3.	Does it provide theoretical clarification on phenomenon under investigation?	
4.	Does it illuminate the relationships among theoretical constructs/ concepts?	
5.	Does it clarify to the readers, what the study seeks to achieve and how that will be achieved?	
6.	If it is in diagrammatic form, does it make sense and have meaningful symbols and denotations?	

Contd.

7. Is it accompanied by relevant descriptive narration and relevant theories and beliefs that support your presumed relationships ?
8. If you have developed conceptual categories, are these directly tied to the research questions ?
9. Do you have at least one conceptual category for every research question ?
10. Have you included all the relevant descriptors that are based on the literature, pilot studies, and your own hunches ?
11. If you have developed conceptual categories, are these directly tied to the research questions ?

Illustration 1: Conceptual framework *"factors affecting the decision to keep a dependent adult child at home"*

The following factors appear to influence the decisions to keep at home an adult family member who is dependent because of disabilities, rather than "placing" or 'institutionalizing' the adult child.

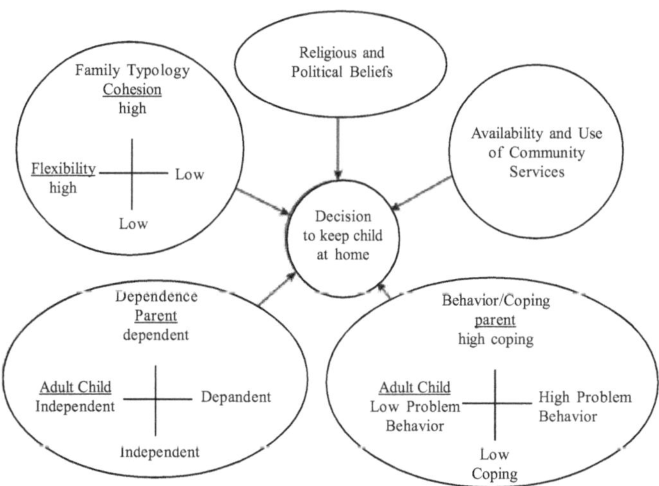

Family Typology is a model of intrafamily interactions and permeability of family boundaries developed by David Kantor and expanded by Larry Constantine. Although data on family typologies has not been collected, intuition and existing data favour the prediction that families in the upper-right quadrant (closed family systems) and lower-right quadrant (synchronous family systems) are more likely to keep the dependent adult child at home, whereas families in the upper-left quadrant (open families) and lower-left quadrant (random families) are more likely to place the adult child.

In the Dependence grid, preliminary data indicate that the upper-left quadrant (high parental dependence, low child dependence) tends to correlate with a decision to keep the adult child at home, whereas the lower-right quadrant (parental independence, high care needs in a child) tends to correlate with placing the adult child. Similarly, in the behaviour/coping grid, the upper-left quadrant (minimal behaviour problems, high parental coping) tends to correlate with keeping the adult child at home, whereas the lower-right quadrant (serious behaviour problems, low parental coping) tends to correlate with a decision to place the adult child.

[*Source*: Adopted from Maxwell, J. (1996). *Qualitative research design: An interactive approach*]

Illustration 2: Conceptual framework for variations in dioxin risk perceptions, behavioural preferences among social groups

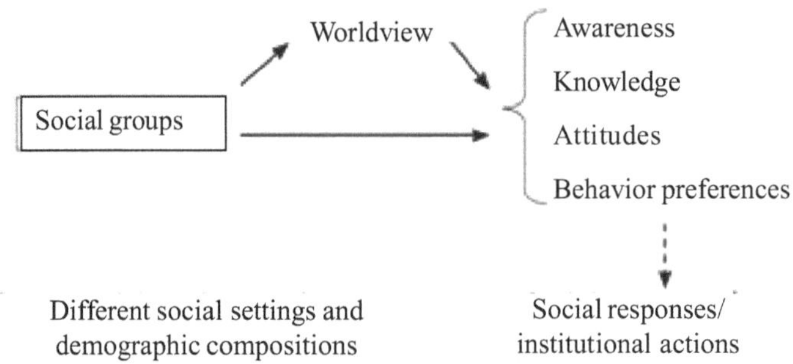

Researchers have examined various factors, in terms of social as well as cultural variables, within different ethnic backgrounds, that influence the variability of risk perceptions and judgments. Although South Korea has a single indigenous ethnic group, the variations among sub-cultural or social groups within the country exist in values or worldview. Literature has identified shared experience, values, and belief that influence subjective risk perceptions within groups This study examined variations among different social group in values, dioxin risk awareness, knowledge, and behavioural preferences. Figure illustrates the research framework which conceptualizes interrelationships among different social group among social settings, values, and risk perceptions among different social groups. According to the framework, the hypotheses developed are as follows:

1. There are differences among social groups in values, awareness, knowledge, and behavioural preferences (willingness-to-act [WTA] for risk reduction behaviours) in relation to dioxin risk.
2. There are factors within group variations influencing dioxin risk reduction behaviours.

[*Source*: [Adapted from Park, S., & Kim, J. G. (2014). *Risk and culture: variations in dioxin risk perceptions, behavioral preferences among social groups in South Korea*, published in *Environmental health and toxicology*]

Exercise

1. Identify any of three conceptual frameworks developed and published in peer reviewed research journals and evaluate them.
2. Develop a diagrammatic conceptual framework for each of the three research problems which you designed in previous section.
3. Provide narrative conceptual framework for each of the diagrammatic conceptual framework you prepared.
4. What do you understand by the statement "building of conceptual framework is an iterative process?"

_ _

_ _

_ _

_ _

_ _

_ _

_ _

_ _

_ _

_ _

_ _

_ _

_ _

_ _

_ _

_ _

_ _

_ _

References

Ravitch, S. and Riggan, M. (2012). *Reason & rigor: How conceptual frameworks guide research.* Thousand Oaks: Sage Publications.

Punch, K. (1998). *Introduction to social research: Quantitative and qualitative approaches.* London: SAGE Publications.

Maxwell, J. (1996). *Qualitative research design: An interactive approach.* Thousand Oaks, Calif.: Sage Publications.

Miles, M., & Huberman, A. (1994). *Qualitative data analysis an expanded sourcebook* (2.nd ed.). Thousand Oaks: SAGE.

Bloomberg, L., and Volpe, M. (2012). *Completing your qualitative dissertation: A road map from beginning to end* (2nd ed.). Thousand Oaks, Calif.: SAGE Publications.

Blaxter, L., Hughes, C. and Tight, M. (2010). *How to research* (4th ed.). Maidenhead: Open University Press/McGraw-Hill Education.

Park, S. and Kim, J. G. (2014). Risk and culture: variations in dioxin risk perceptions, behavioral preferences among social groups in South Korea. *Environmental health and toxicology, 29.*

Chapter 4

Concept of Hypothesis and its Formulation

Most of the scientific research studies –except few descriptive survey research studies–will have one or more hypotheses to be tested as a part of study. *Hypothesis is a prediction intended to be tested in a research study* (Aron, Coups, and Aron, 2012). Hypothesis is the guide which directs the purpose and process of research by indicating the dependent and independent variables to be observed, extraneous variables to be controlled and the active variables to be manipulated.

Hypothesis is deduced from the scientific problem. Suppose our research problem is stated as "What is the impact of training programme on the effectiveness of job performance VLWs". Then our research hypothesis will be "*training has significantly enhanced the job performance of VLWs*" OR "training has no significant impact on the job performance of VLWs".

Note that research problem should not be confused with research hypothesis. Hypothesis is relatively more specific and empirically testable.

Difference between Problem and Hypothesis

Sl. No.	Problem	Hypothesis
1	Relatively more general and abstract	More specific and highly operationalized
2	Problem cannot be tested	Hypothesis can be tested
3	It is stated in interrogative form	Stated in declarative form
4	It asks relationship between two or more variables	It assumes relationship between two or more variables
5	As a consequence of research, we solve problem	As a consequence of testing hypothesis, we reject/ accept null/alternate hypothesis

Importance of Hypothesis (Role of hypothesis)

a. *Provides point of enquiry to research*: Hypothesis specifies the purpose of research and makes the research problem amenable to statistical testing.

b. *Directs the process of research*: Hypothesis directs the process of research by specifying what variables should be operationalized, observed, measured, compared and contrasted.

c. *Specify the type of observation to be made and data to be collected*: Looking at the hypothesis one can easily understand the nature of variables under investigation (quantitative, qualitative) and accordingly design the observation methods and procedures for data collection.

d. *Helps in drawing conclusion*: Hypothesis provides an overview of possible tentative conclusions that can be made from any given research study.

e. *Tool of advancement of knowledge*: Hypothesis connects the theory and observation and acts as tool for advancement of theoretical knowledge base.

Types of hypothesis: Hypothesis can be classified into number of ways, but in the context of social science, below given two classifications are of greater relevance.

Based on the existence of difference relationship between parameter values

Null hypothesis (H_0): Null hypothesis argues that variations in sample observations are matter of pure chance. This is the hypothesis of no difference. In this hypothesis it is claimed that, there exists no significant difference relationship between the sample parameter (mean, standard deviation, variance etc.) values. Alternatively, it may also claim that, there exists no significant difference between observed and expected parameter values. Null hypothesis is denoted as $H_{0.}$

For example, suppose we want to introduce a training method "X" to impart agricultural job skills among farmers. Prior to that, we want to test whether this training method have significant impact of job skill competencies. Now our research problem will be "What is the impact of "X" training on the agricultural job skill competency of the farmers?"

The corresponding null hypothesis will be "X Training has no significant impact on the agricultural job skill competency of the farmers". This null hypothesis can be also written as "there is no significant difference between agricultural job skill competencies of the farmers, before and after the training"

Alternate hypothesis (H_1): An alternate hypothesis is the hypothesis which differs from the hypothesis being tested (Rich, 2003). It is the hypothesis of significant difference or significant relationship. It claims that sample observations are influenced by some non-random cause (in case of relational hypothesis). Alternate hypothesis is denoted as H_1. For the previously stated research problem, the corresponding alternate hypothesis will be "X Training has significant impact on the agricultural job skill competency of the farmers."

Note that, in this type of alternate hypothesis we are simply claiming that there exists a significant difference between the pre and post training Job skill competencies; but we are not stating, whether the job skill competency has significantly increased or decreased after the training. This type of alternate hypothesis which does not specify the direction of difference change is known as "*Non-directional hypothesis.*"

On the other hand, if we state the alternate hypothesis specifying the direction of change, then it will be "*directional hypothesis*". The directional hypothesis for the previous example would be "*X training has significantly increased the agricultural job skill competency of the farmers.*"

Alternatively the another possible directional hypothesis can be *"X training has significantly reduced the agricultural job skill competency of the farmers."*

Note that, we always test and reject the null hypothesis and not the alternate hypothesis.

Based on the nature of the study

1. *Relational hypothesis:* It assumes existence or non-existence of relationship among two variables. This type of hypothesis are formulated for the studies that investigate the causal relationships.

   Example: There exists a significant relationship between amount of fertilizer applied and the crop yield.

2. *Descriptive hypothesis:* It makes tentative proposition about existence, size, form and distribution of a particular natural phenomenon. Such type of hypothesis are formulated for descriptive studies.

   Example:

   - Innovative farmers are relatively younger than the conservative traditional farmers.
   - Paddy productivity was relatively higher in case of small farmers as compared to large farmers.

Characteristics of good hypothesis

1. It should be simple
2. Should not be too general or too specific
3. Conceptually clear: It should use existing terminology, if new term coined, it should be operationally defined in generally known terms
4. It should be related to current body of knowledge
5. It should be amenable to the empirical testing
6. It should be in line with available techniques of observation, measurement and analysis
7. It should be free from value judgment.

Sources of Hypothesis

1. *General culture:* The society in which an individual lives, the pattern of behavior among the society members, value system of the given society etc. can be a source for formulating the hypothesis in the field of sociology, education, psychology and other fields of social science.
2. *Experience:* Experience of working in particular field of research makes it possible to understand the current level of advancement made in the particular field of research and understand the existing practical lacuna and knowledge gaps. This understanding of knowledge gap is an important source for formulating both exploratory and relational hypotheses.
3. *Review of literature:* Understanding of research work done in past provides as insight into the current level of knowledge advancement and existing research procedures. Through analysis and synthesis of such review studies, one can formulate the hypothesis for future research work.
4. *Specific theory:* Careful evaluation of well established theories, sometimes may lead to discovery of limitations in the theories, conditions where the theory doesn't operate or the methodological loopholes in testing of the theory. Under such situations one can make an attempt to improvise the theory, by using the relatively more advanced methodologies or by including new constructs in the theory or by eliminating the existing constructs or by improvising the definition of constructs and then retesting the hypothesis to prove/disprove the theory.
5. *Analogy*: One can draw new hypothesis from related hypothesis which were tested in past. We can add new independent variables or modify

constructs and refine the hypothesis to formulate scientifically improved new hypothesis from related and previously tested hypothesis.

Exercise

a. Select any 5 research papers published in peer reviewed journals and identify which type of hypotheses the researcher has used (relational, descriptive / directional, non directional)

b. Based on the research problem which you developed in previous section, formulate the corresponding null and alternate hypotheses

c. Is it necessary that every research must have some hypothesis? Justify your answer with example

Reference

Aron, A. and Coups, E. (2013). *Statistics for psychology* (6th ed.). Boston: Pearson.

Rich, P. (2003). *Understanding, assessing, and rehabilitating juvenile sexual offenders.* Hoboken, N.J.: John Wiley & Sons.

Chapter 5

Variables: Concept, Classification and Operationalization

Variable is any character or attribute of the given object/set of objects which varies from object to object or from one point of time to another point of time. For example height of individuals is a variable; it varies from person to person. Kerlinger (1973) defined variable as a *symbol to which numerals or values are assigned.*

Importance of variables: In previous sections, we discussed about formulating the research problem and development of research hypothesis. Variables are the building blocks of any scientific research problem and research hypothesis. When we consider a hypothesis to be relational hypothesis, we mean that it make claims about relationship between two or more variables. When we test hypothesis, as such we do not test the hypothesis, rather we test the significance of hypothesised relationship between two or more variables. Since variables play central role in observation, measurement and testing of relationship, it is of prime importance to understand the meaning of variables, different types and the way we define these variables.

Criteria for selecting research variables

i. Variable should be related to body of science and should not be based on metaphysical descriptions

ii. Variable should be either a presumed cause, effect or a significant extraneous variable

iii. It should be measurable directly or indirectly

iv. Selected variable should be relevant to the research problem

v. Manipulation and measurement of variable should be economically feasible and possible in the given time limits.

Typology of variables

There are numerous classifications of variables existing in the field of experimentation. Considering the present context of this manual, we will limit our discussion only to those kinds of variables which make sense in developing research design, rather than considering the aspect of data analytics.

Based on the cause-effect relationship

Independent variable: It is the presumed cause of the dependent variable. It is also called as antecedent variable. We can say that, any variable which is the cause of variation in the dependent variable is the independent variable.

Dependent variable: The variable that varies with variation in independent variable. It is the consequent variable.

Note that a variable may act as dependent variable in one study and same may be the independent variable in some other study. Vice versa is also true. This classification of variables is not an absolute classification. Independent and dependent variable are relative terms.

Based on nature of values taken by variable

1. *Categorical:* In this case, the subjects are grouped into categories based on whether the given subject possess an attribute/character or not (nominal). Alternatively the subjects may be grouped into various categories based on the amount/extent of given attribute/character found to be associated with given subjects (ordinal categories) Example: Students can be classified into meritorious, average and underperformers based on academic marks range.

    a. *Nominal*: Here the Observations can take values which cannot be organised in a logical sequence of ascending or descending order. These variables represent the character which can be used for group ing of the objects. Example: Sex, Marital status, Family type etc. Note that here assigned numbers to each category do not have any numeric meaning. They are just symbols. Nominal variables are measured at nominal level of measurement.

    b. *Ordinal*: This is also a type of variable which can be used to group the objects into categories, which can be rank ordered. For example academic grades (A, B, C) and attitude (highly negative, negative, positive and highly positive) are ordinal variables. Here assigned numbers do not have absolute meaning associated with them but they have meaning to the extent that they can be rank ordered.

2. *Numeric:* These variables are measured at interval/ratio level. Here assigned numeral have certain kind of meaning. For example if first object is assigned value 1 and second object is assigned value 2 then we can atleast rank order these variables and we can calculate absolute difference between the values of two objects.

   a. *Discrete*: Variable having only integer values (1, 2, 3 etc.). For example, No. of children, number of trainings conducted, family size etc. Discrete variables are generally qualitative in nature.

   b. *Continuous*: Observations can take any value between a certain set of real numbers (*e.g.*, height of a plant may be recorded something like this; 2.546 cm). The value given to an observation for a continuous variable can include values as small as the instrument of measurement allows. A continuous variable may take both fraction and discrete values. Examples of continuous variables are height, time, age, and temperature.

Independent variables of practical importance in research

Extraneous variable: Independent variable that causes variation in dependent variable whose effect the researcher wishes to control, because it is not the independent variable of interest to the researcher.

For Example: When a researcher is interested in investigating the effect of income level on academic performance of the students, parents' educational level is an independent intervening variable, because it also affects the students' academic performance. But it is not the variable of interest to the researcher and hence it is an extraneous variable whose effect researcher wishes to control.

Intervening variable: A variable that explains a relation or provides a causal link between other variables. Also called by some authors as "mediating variable" or "intermediary variable". Intervening variable is a control variable that follows an independent variable but precedes the dependent variable in a causal sequence.

Example: The statistical relationship between farmers' income and his cosmopoliteness needs to be explained because just having money does not make one cosmopolite. Other variables intervene between money and cosmopoliteness. Farmers' with high income tend to have better access to ICT tools than those with low incomes. Here access to ICT tools is an intervening variable. It mediates the relation between farmers' income and his cosmopoliteness.

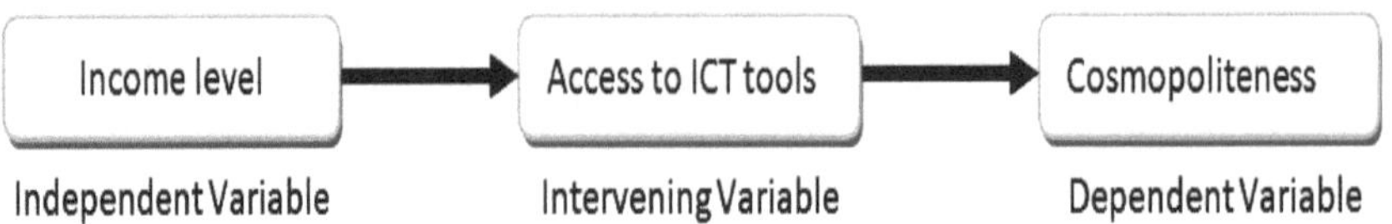

Latent variable: It is an underlying variable which cannot be observed and measured directly. But it is interpreted from the other observable variables which are used as indicators. For example we cannot observe and measure intelligence directly; hence we observe one's vocabulary, IQ level, professional success etc. to interpret about his/her intelligence level. Latent variables are also called as constructs.

Manifest variable: It is the variable which can be observed and measured directly and is used as an indicator for the presence of a latent variable. It is also known as an indicator variable. For example, we cannot observe and measure the intelligence directly (hence it is a latent variable). Instead we will interpret the level of intelligence from indicators such as vocabulary size, IQ test score, ability to play complicated games well and so on. The goal is always to achieve convergence between manifest and corresponding latent variables to the extent possible.

Moderating variable: It is a variable which moderates (changes) the strength or direction of relationship between two variables. For example, work stress increases drinking problems for people with a highly avoidant coping style, but work stress is not related to drinking problems for people who score low on avoidant coping (Cooper, Russell, & Frone, 1990). Moderators indicate when or under what conditions a particular effect can be expected. The effect of a moderating variable is characterized statistically as an interaction; that is categorical.

Confounding variable: It is an independent variable that confounds the effects of another independent variable. Suppose, a researcher is interested in deciding which one of the two given training programmes (X and Y) is more effective in enhancing financial management skills among farmers. He randomly selects two villages (A and B). Now he administers training programme X and Y to villages A and B respectively. Consequently he found that, there was relatively higher increment in farmers' financial management skills in village A. He interprets that, training A is more effective than training B. Preliminary survey found that, farmers in village A are economically more prosperous and have relatively higher educational qualification than compared to village B. There is no way to determine if differences in enhancement of financial skills between the two villagers are caused either by difference in educational qualification or by differences in training programme or by both of the independent variables.

Concept, Construct, Definition

We cannot use the variables in research framework and hypothesis testing as such. We need to refine, define and transform the variable into manageable, observable, measurable and statistically testable form. Prior to that, we need to understand the meaning of certain underlying aspects like concept, construct and definition of variables.

Concept is the expression of abstraction formed by generalization from particulars (Kerlinger, 1973). We define the concept by bringing together various observed behaviors expressed in terms of words. For example, we may club together the observed behaviors on the part of farmers –curiosity to try new things, urge to meet standards of excellence, risk taking etc. –into a single concept of "achievement motivation".

Whenever we develop research based theoretical scheme, rather than concept, it is the construct that enters into the theoretical scheme. Concept and construct both seem to be similar, both are closely related but they do differ sharply. *Construct is a concept which had added meaning having been deliberately and consciously adopted for specific research.* Concept is more general whereas construct research context specific. Concept is more broad whereas construct is more specific and do not enter into scientific hypothesis testing whereas construct enters into statement of scientific hypothesis. Researcher tests the relationship between two constructs and not between two concepts.

Note that, constructs are not directly observable and measurable. Hence we need to redefine these constitutively defined constructs into observable and measurable operationalized variables. In order to do this, first we need to understand the meaning of and difference between constitutive and operational definitions of the construct.

	Constitutive definition	Operational definition
Meaning	Defining a construct by using other constructs	Defining a construct in terms of how it is manipulated or how it is observed and measured
Scope	More broad and universal	More specific and defined in context of particular research
Content	Doesn't specify how manipulation was done or how to observe the construct and measure it	Manipulation, observation and measurement of construct are specified
Testing	Have least implications for statistical testing	Have larger implications for statistical testing
Example	Anxiety is defined as subjective fear	Anxiety is operationalized as score obtained by the respondent in anxiety test

Operational definition: An operational definition assigns meaning to a construct or a variable either by specifying the activities or operations necessary to measure it or by specifying the researchers' manipulation of it (Reference). There are two types of operational definitions.

Experimental Operational definition: This definition describes the researchers' manipulation of a variable. For example the construct 'reinforcement' can be operationalized by specifying how the desirable response of the subject must be awarded.

Measured Operational definition: This Operationalization describes how a variable will be observed and measured. For example achievement motivation may be operationally defined as score obtained in standardized achievement test.

Importance of Operational definitions

a. O.D. Lays down the guidelines for observation and measurement of a variable. Thus, they reduce subjective bias and errors in observation and measurement. Consequently, it improves the objectivity of the research.

b. Operational definition clarifies both readers and future researchers on how manipulation was carried out in research. Thus, it helps in identifying possible loopholes in the manipulation method and provide an opportunity to re-test the hypothesis by refining manipulation method.

c. Operational definition act as bridge between Theory-Hypothesis-Construct and observation level.

Exercise

1. Review the literature on the three research problems of your interest and list out at least three variables (which were operationally defined) in each of the three research publications.
2. Identify 3 major constructs that emerge in your previously developed research problem. Further, write down the constitutive and operational (both experimental and measured) definition of the identified constructs.
3. Enlist three examples of below given types of variables.
   i. Discrete ordinal variable
   ii. Continuous ordinal variable
   iii. Nominal variable
   iv. Dummy variable.
4. Classification of independent and dependent variables is a relative classification. Justify this statement with example.

References

Cooper, M., Russell, M. and Frone, M. (1990). Work Stress and Alcohol Effects: A Test of Stress-Induced Drinking. *Journal of Health and Social Behavior,* 260-260.

Kerlinger, F. (1973). *Foundations of behavioural research* (2nd ed.). New York: Holt, Rinehart and Winston.

Chapter 6

Levels of Measurement

There can be no scientific studies which do not involve the process of measuring. Measurement process is the integral part of research activity. Measurement can be defined as the *assignment of numerals to the objects or events according to set of rules*. The emphasis in this definition is on phrase '*set of rules'*. If the rules can be set up based on some rational or empirical basis, then theoretically anything can be measured. It is very much easy to understand the concept of measurement in case of physical sciences (Example weight, length, mass etc.) but it is harder understand the concept of measurement in case of psychological objects/phenomenon (intelligence, aggressiveness, attitude etc.) without understanding the predetermined rules of measurement. A rule is a guide, command that tells you what to do/how to do.

When we say measurement is the assignment of numerals to the objects/events then it means that assigned number should either implicitly indicate something or should have explicit numerical meaning as in mathematics. When in our day to day life, we are saying that friend X is more intelligent than friend Y then also we are measuring them, but based on some rules. The rule here may be "judge the intelligence level based on academic performance, higher the academic performance, higher the intelligence level of person".

Mere measurement of the variable is not the goal of the research process. It is just an intermediate step in reaching ultimate goal of discovering underlying relationships and predicting the phenomenon. In order to do this, measured data need to be analysed by using the statistical methods. Statistics–like other fields of mathematics–highly relies on rationale and logical thinking. So, to get more accurate results of analysis, we need to measure the variable and collect the data that suits the statistical procedures and minimise the cost of research work. To collect the data that is suitable for particular statistical procedure, one must develop data collection/measurement tools that suit the requirement. In social research, one must have adequate knowledge about the levels of measurement and amenability of various levels of data for different statistical procedures, in order to get results with minimum error and high predictability.

Levels of measurement: Based on the meaningfulness of assigned numerals and existence of absolute zero in the measurement scale, the measurements are classified into 4 levels. These are hierarchically arranged in terms of their superiority. Most superior level of measurement is one which have absolute zero and least desirable measurement is the one which do not even possess any meaning to the assigned numeral, this so because the data generated from measurement with absolute zero is amenable to every statistical procedure and as we move to the lower levels of measurement the amenability of data to statistical procedure become more and more limited.

Before studying the levels of measurement one must understand the postulates of measurement. Postulates guide in deciding the level of measurement. Before making any conclusion based on postulates, always keep in mind that the basic rule of categorisation, especially in nominal level of measurement (discussed later) is that categories should be exclusive. That means when you are categorizing the objects, categories, should not overlap with each other.

Postulates of measurement: There are four postulates of measurement. A postulate is an assumption, an essential prerequisite for carrying out some operations or line of thinking. Here in this case, it is an assumption of the relation between the objects being measured. Below given are the four postulates of measurement.

1. *Either a = b or a = b, but not both : Classification postulate*

Either 'a' is equal to 'b' or 'a' is not equal to 'b' but not both. This postulate is necessary for classification. This rule helps in avoiding overlapping of categories. Given the two objects, they can be either same or differ with respect to amount of a single point character but both cannot be equal and not equal at a time with respect to amount of certain character possessed by them. Here equal refers not to exactly equal but sufficiently equal.

2. *If a = b, b = c then a = c : Equivalence postulate*

This means if one member of the universe is same as the second member, and second member is the same as the third member; then first member is same as the third member. This postulate is necessary to establish equality among the members assigned into single category. For example, when we categorise the respondents into male and female then we assume that all the members in the category "Male" are equal with respect to their maleness.

3. *If a > b and b > c then a > c* : Ordinal *Transitivity postulate*

This postulate indicates that, if any object 'a' possesses character 'X' in the amount greater than that of possessed by object 'b'; and item 'b' possesses same character more than what object 'c' possesses, then we can conclude that item 'a' possesses character 'X' in the quantity greater than object 'c'. This postulate is necessary for rank ordering of the objects.

4. if a = xb then a –xb = 0 : Absolute Zero Postulate

Note here, this is most important essentiality to call a measurement; the ratio level measurement. When we say that person X weighs 50kg and person Y weighs 100 kg, then we can also say that person Y weighs double the weigh to person X. Practically the difference between weight of two persons weighing 50 Kg and 100 kg is zero. But this is not true when a scale operates at the interval level.

Levels of Measurement: Based on the ability of measured values to satisfy the above said postulates measurements are classified into four levels.

1. Nominal Level of Measurement
2. Ordinal Level of Measurement
3. Interval Level of Measurement
4. Ratio Level of Measurements

Nominal Level of Measurement: It is the lowest level of measurement. Here the assigned numeral to the object does not have any meaning. Since it do not have mathematical meaning, we cannot rank order the objects based on the assigned numeral. They are much like labels. For example, when we are conducting survey we may code Male as 1 and Female as 2. This really doesn't mean females measure high on gender measurement. We cannot rank order the respondents based on gender. We cannot say that attribute gender is highly possessed by females and less possessed by males. Other examples of such measurement level include Religion, Race, and Political View *etc*. Can we rank order race or religion of the individual? Of course, we cannot. This level of measurement lacks meaning to the assigned number and do not have either true or relative zero, hence this is the lowest level of measurement. Based on the above discussion we can say that any measurement which follows *classification postulate and equivalence postulate but do not follow the transitivity postulate and absolute zero postulate is the Nominal Measurement.*

Only counting is the possible mathematical operation in this case, so such kind of data is amenable only to simple statistical procedures like Frequency,

Percentage and Chi-Square testing etc. To suit such type of data generally non-parametric statistical methods are used, there is limited opportunity to use parametric methods with such kind of data.

Ordinal Level of Measurement: As the name suggests, in this case the assigned numerals can be rank ordered since they have certain mathematical meaning associated with them.Example: GPA (Grade Point Average) of the students. We can rank order the students based on theirGPA because the GPA numerals have certain mathematical meaning, they reflect the level of academic performance. If there are three students —X, Y and Z with merit rank I,II, III respectively, then we can order them based on their academic performance (Y > Z > X) looking at their ranks, so here rank order is the measure of academic performance and ranks have implicit meanings associated with them. We can definitely rank order them but, if three students have rank I, II, III then does it mean the difference in academic performance of rank I and II is same as difference between performance of rank II and III? Can we tell that difference between rank I student and Rank III student is twice the difference between rank I and II? Of course we cannot say that. An ordinal level of measurement lacks absolute zero, which makes it impossible to compare the objects in terms of multiples. So it fails to satisfy the postulate of absolute zero.

So, it can be concluded that any *measurement that satisfies the transitivity postulate but the distance between any two consecutive ranks is not equal and lacks absolute zero, then it is ordinal level of measurement.*

Interval Level of Measurement: Equal /equal-interval scale possess all the characters of an ordinal scale and additionally the distance between any two consecutive intervals is equal. So the intervals here can be added and subtracted. In other words addition and subtraction are the additional possible mathematical procedures in this level of measurement. Classical example of this type of scale is Celsius temperature scale. The difference between 30 degrees and 31 degrees represents the same temperature difference as the difference between 31 degrees and 32 degrees. This is because each 1-degree interval has the same physical meaning (in terms of the kinetic energy of molecules), but we cannot say that when temperature is 90^0 C then molecules have thrice the kinetic energy as that of molecules at 30^0C, because it also accounts for the kinetic energy of the molecules at the temperatures below 0^0C.

Here in this case, the distance between any two consecutive intervals is same but it lacks absolute zero. When temperature is zero on Celsius scale, it doesn't mean temperature cease to exist. Additionally, if the average temperature in City A is 18^0 C and average temperature in city B is 36^0 C, then technically we cannot say that temperature in city B is twice the temperature in City A. So, *any level of measurement that follows the transitivity postulate but lacks absolute zero is interval level of measurement.*

Here the measured values are relative values but the difference between two values is absolute. Hence we can add and subtract the values but we cannot divide two values obtained measured at interval level but we can divide the differences between two values obtained by using interval scale.

Ratio Level of Measurement: This is the highest level of measurement because it satisfies transitivity law. Distance between any two consecutive intervals is equal and this type of scale always has absolute zero hence amenable to wide range of statistical procedures that require multiplication and division. It is the most desired level of measurement. Addition, subtraction, multiplication and division are possible mathematical operations in the ratio level of measurement. Length, mass, income, age etc. are the examples of ratio level of measurement.

In social science, most of the measurements are at ordinal level but we assume them to operate at interval level. Ratio levels of measurements are commonly found in physical sciences.

Level of measurement and Mathematical operations

Level	(*f*)	Ranking	(+)	(-)	(*X*)	(÷)
Nominal	√	X	X	X	X	X
Ordinal	√	√	X	X	X	X
Interval	√	√	√	√	X	X
Ratio	√	√	√	√	√	√

Note: (f) = Counting, (+) = Addition, (-) = Subtraction, (X) = Multiplication, (÷) = Division, √ = Possible,, X = Not possible

Exercise

Evaluate the below given short case studies and answer the corresponding questions:

1. A researcher is interested in comparing the intelligence level of the two students from two different schools. For this, he develops a test to measure the intelligence level of the students and administers the test to both the students. Student from school A secured 35 marks whereas the student from school B secured 70 marks in the intelligence test. The teacher interprets that "student B is two times more intelligent than the student A". Is he correct in his interpretation? Justify your answer.

2. A researcher is curious on knowing the relationship between gender and the intelligence level and uses correlation coefficient to explore the relationship. Is he right in his approach? Justify your answer.
3. Why ratio level of measurement is most preferred level of measurement?
4. A researcher is interested in investigating the impact of gender on entrepreneurial success among the farmers. He uses correlation coefficient as a measure of relationship between gender and income level. He obtained a correlation coefficient 0.45. Further he found that this relationship is statistically significant at 5% level of significance. Comment on this research approach in the context of measurement level and test statistics.

Reference

Kerlinger, F. (1973). *Foundations of Behavioural Research* (2nd ed.). New York: Holt, Rinehart and Winston.

Chapter 7

Concept of Reliability

Reliability deals with the dependency, consistency, stability, predictability and accuracy of the research data. In our day to day life, when we say that a person is reliable then it means his behaviour is predictable, his behaviour (the way he behaved in past, he is behaving now and the way he will behave in future) will be consistent and he is dependable. On other hand an unreliable person's behaviour is unpredictably variable and he cannot be depended upon.

Similarly, in case of social research the measuring instrument may be either relatively reliable or unreliable. Hence prior to using of an instrument for measuring the variables under social research, we must ensure that measuring instrument *per se* is reliable. Before quantifying the reliability of the measuring instrument, we must first understand the concept of reliability and the way it is defined in the context of science and the importance of reliability. Conventionally there are two approaches for defining reliability. First approach of defining reliability will be by asking the question.

"***If we measure the same set of objects repeatedly by using same or comparable measuring instrument, will we get the similar measures?***"

This aspect of reliability covers the definition of reliability in terms of dependency, consistency, stability, predictability.

Second approach to defining reliability can be summarized by asking the question ***"Whether the measures obtained from a measuring instrument are the true measures of the property being measured?"*** In other words under this approach of defining reliability we want to know to what extent the measured values represent the true values? This is the *accuracy* approach to defining the reliability.

Characters of reliability

a. It is the measure of consistency, predictability, reproducibility, stability, accuracy and dependability of the test scores

b. Estimated reliability is a matter of degree

c. It is the character of test scores/results and not the character of measuring instrument itself

d. It doesn't tell anything about, to what extent we are measuring, what we are supposed to measure

e. Reliability is dependent on characteristics of the test, the conditions of administration, and the group of examinees

f. It can only be estimated.

Theory of reliability

Theory of reliability is based on the *true score theory*. It is not exactly a theory, rather it is a postulate. It states that *measured score* is additive composite of the two components –*true ability score* and *the error*.

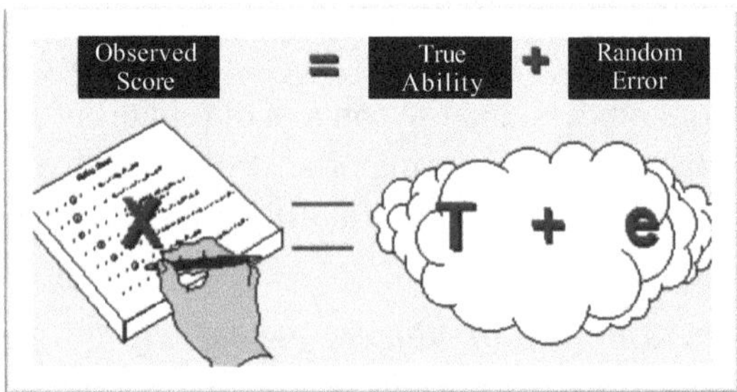

Mathematically this can be written in the form of equation,

$$T_x = T_\infty + T_{e,} \qquad \text{Equation I}$$

Where, T_x is total score, T_∞ is true score which can be conceived to be the score an individual would obtain if all internal and external conditions were perfect and the measuring instrument were perfect and T_e is the error score.

Suppose, the test is administered to same individual repeatedly for large number of times, then we can calculate the variance for these set of scores. Now we can extend the *equation I* to the variance of the set of score. This relation can be expressed mathematically as

$$V_x = V_\infty + V_{e,} \qquad \text{Equation II}$$

Where, V_x is the total variance, V_∞ is true variance for the attribute under investigation and V_e is the error variance.

Now we can define reliability of the measuring instrument in terms of error variance and true score variance. We can say that *greater is the relative size of true variance, higher is the instrument's reliability. Greater the error variance, lower is the reliability of the measuring instrument.* Further we can define reliability in ratio terms

Reliability is the ratio of true variance to the total variance

$$r_{tt} = \frac{V_\infty}{V_t}$$

Alternatively we can also say that, Reliability is the ratio of error variance to the total variance, subtracted from unity

$$r_{tt} = 1 - \frac{V_e}{V_t},$$

It is same as,

$$r_{tt} = \frac{V_t - V_e}{V_t}$$

Hence, if we can measure the true variance or the error variance then we can also measure the reliability of the measuring instrument. But, the real problem is that *we can neither measure the error variance nor the true variance.* So where does it lead us to then?

We cannot measure the true or error variance; but we can estimate the error variance and through this we can also estimate the reliability of measuring instrument.

In summary, reliability is the accuracy and consistency with which it measures whatsoever it measures. Reliability is not related to whether the instrument is measuring what it was supposed to measure.

Exercise

I. Judge whether following statements are true or false and justify your answer.

1. A measure that has no random error is perfectly reliable.

2. It is possible to calculate absolute reliability of the measuring instrument.
3. True score of an individual is always constant, while error scores vary randomly.
4. It is possible to calculate the reliability.
5. Reliability is the property of measuring instrument.

II. **Case Study:** Suppose a veteran shooter is testing four guns for their reliability. He fixes guns on the perfectly stable platform and aims the guns at the bull's eye of the target. Below are the images of the target after ten rounds of shooting. Rank order the guns in the ascending order of their reliability.

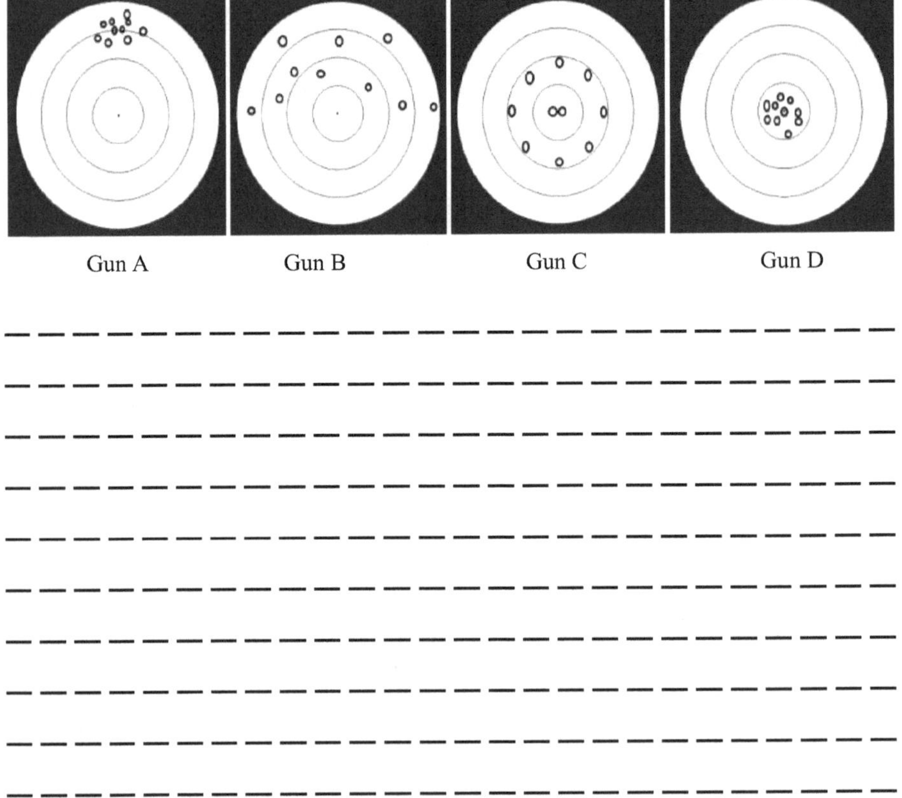

Gun A Gun B Gun C Gun D

Reference

Kerlinger, F. (1973). *Foundations of Behavioural Research* (2nd ed.). New York: Holt, Rinehart and Winston.

Chapter 8

Measurement of Reliability

In previous section, we discussed about the concept of the reliability and its importance. In this section, we will discuss about different types of reliability and procedures for estimation of the reliability. We already discussed that one cannot measure the reliability but of course it can be estimated.

Various authors follow different nomenclature for classifying reliability. Based on review of wide variety of literature, we classified the reliability into four types, namely:

i. Test-Retest reliability

ii. Inter-Rater reliability

iii. Parallel-Forms reliability

iv. Internal Consistency reliability

In all these, the common aspect is the comparison of measurements at two different time points/ with two different individuals/ between two set of individuals/ with two different forms of stimuli. We will discuss in detail each of these four measures of reliability.

Test-Retest reliability

Test-retest reliability aims at estimating the measuring instrument's temporal stability. In other words, we will measure that, how much stable the measurement scores are over a period of time. In this method, we will administer the same test to same individuals but at different occasions. For a test to be reliable,sets of scores must exhibit high inter-correlation.

***Practical limitations of the test-retest reliability*:** One major limitation of the test retest reliability is that if time gap is too short then this method suffers from memory effect, confidence effect and practice effects. Majority of us will remember our responses (in first test) and when we start to respond for second time to the same test, we tend to respond in the way we did on the first attempt rather than going through the questions once again and rethinking. This creates an artificially high reliability coefficient as subjects respond based on memory of previous test, rather than the rethinking. On other hand, when the time gap is too long then this method suffers from maturity effect. Owing to difficulties in controlling extraneous influencing factors along with memory effects, carry over effect, confidence effects and practice effect, the test-retest method is generally less useful as compared to other methods.

Procedure

i. Administer the test/scale to the given group of individuals

ii. Record the observed/measured scores

iii. After a specific time interval –preferably after 15 days –administer the same test/scale to same group of individuals

iv. Once again, Record the observed/measured scores

v. Now correlate the Scores of each individual obtained at two different time points.

Conclusion: Greater the inter-correlation between two set of scores, higher is the reliability of measuring instrument.

Inter-Rater reliability

As the name itself suggests, we will correlate here the ratings given by two observers for a single object or phenomenon. The question arising at this stage is why the inter-rater reliability is required? Inter-rater reliability is required because we involve humans as a part of measurement process, especially we involve human for observing social / psychological phenomenon and rate the intensity of phenomenon. Ensuring of inter-rater reliability is required because human being have limitation of being biased and in consistent observing, judging. Interpretation and judgement are influenced by the emotions and subjective perceptions of the judge and more importantly we human beings loose efficiency and get tired of doing repetitive tasks. This necessitates the estimation of Inter-rater reliability, especially when measurement is based entirely on observation and interpretation of the phenomenon/object by the respondent/expert judge.

Procedure

There are two approaches for calculating inter-rater reliability. First one is when measurement consists of a continuous data and second one is when measurement consists of categories.

- When measurement consists of categories

  a. Administer to the judges, the phenomenon/objects to be rated and ask them to assign the objects across predefined categories (*Example:* Worst, Bad, Good, Better, Best).

  b. Count the number of objects for which both the raters have assigned same category.

  c. Now, Calculate inter-rater reliability by using Cohen's Kappa or Fleiss' Kappa.

  d. When rating consists of categories and there are two raters, then the Cohen's Kappa is a suitable measure of reliability. In case of more than two raters Fleiss' Kappa is more suitable.

Simplest measure of reliability in this case is presented in following formula.

$$\text{Reliability} = \frac{\text{Number of objects assigned same category by both judges}}{\text{Total number of objects}} \text{X}100$$

Note: *E*nsure that the raters are homogenous in experience of the object and are experts in rating the object under investigation

- *When measurement consists of continuous points*

This is the case when the rater is free to rate the objects by using continuous data points rather than assigning the objects to the predefined rating categories. In this case Pearson's r or Spearman's ρ can be used to measure pairwise inter-rater reliability. In case of more than two raters, we will go for mean of correlation coefficient of all possible pairs of ratings.

Parallel-Forms reliability

One way to assure that memory effects do not occur is to use the parallel-forms reliability. Parallel-forms reliability estimates the consistency of the results of two parallel forms of same test constructed in the same way from the same content domain. This is similar to split half method except that here two subsets used are parallel forms which are constructed so that the two forms can be used independent of each other and considered equivalent measures whereas in case of split half method rather than assuming the measures to be equivalent, we aim at estimating the equivalence (homogeneity) of two sets.

Practical limitations of Parallel-Forms Reliability: In this case the memory, practice and carryover effects are minimised but not controlled completely. Here when the two test forms are not truly equivalent, the possibility of making erroneous decision about reliability is greater. Further it demands lot of skill and experience for developing truly parallel forms of single test.

Internal Consistency reliability

As the name suggests this kind of reliability aims at estimating to what extent the items within a test are homogenous. Most common method used to assess internal consistency reliability is the ***Split half Method***. Other commonly used methods in estimating the internal consistency reliability are Cronbach's alpha, the Kuder-Richardson 20 and Hoyt's analysis of variance. Kuder and Richardson Formula 20 is suitable when the responses are dichotomous (Example Yes/No), whereas the Cronbach's alpha can be used with both dichotomous and non-dichotomous (continuous) measures. It was first named alpha by Lee Cronbach in 1951. Cronbach's alpha and Hoyt's analysis of variance give identical results when all items are measured dichotomously.

Split half method Procedure

a. Divide the items within a test into two halves, most of times we use method of splitting the items into two parts wherein odd numbered items (1,3,5...) are grouped into one halve and the even numbered items (2,4,6...) are placed into the other halve.

b. Give clear instructions to the respondents about how to respond to the test items.

c. Administer the both subsets of items (prepared in step 1) simultaneously to the single set of respondents.

d. Inter-correlate the scores obtained by the respondents for two subsets of test items, by using Spearman-Brown formula.

e. Estimate the reliability using the Spearman-Brown Coefficient (ρ).

Note: Rulon/Guttman's formula or Flanagan Formula can also be used for estimating the internal consistency reliability under Split half approach.

$\rho = \frac{2r}{1+r}$, Where,

$\rho = Spearman - Brown\ Coefficient$

r = Inter-correlation between two sets of scores

***Interpretation*:** Higher the value of positive correlation, greater the reliability.

Practical suitability of split half method: One advantage of split half reliability is that it overcomes the variability produced by administration of test at two different time points –i.e. it eliminates the variability produced due to influence of extraneous or intrinsic variables that operate over a period of time. Major advantage of split half method is that it requires single administration of the test in contrary to the repeated administration of test in case of test-retest reliability. But it is not a robust method. Additionally split half method eliminates the carryover effect, memory effect, confidence effect, effect of maturity and boredom which are inherent to test-retest method. Major disadvantage associated with split half method is that it cannot be used in power tests and heterogeneous tests. Additionally this method cannot be used for estimating reliability of speed tests.

Method of rational equivalence: This is another method for estimating the internal consistency reliability of the measuring instrument. In this method we work out the inter-correlation of the items of the test and correlations of each item with all the items of the test. Here it is assumed that all items are equally difficult, the correlation between the items is equal and test is unidimensional and is homogeneous in nature. Kuder and Richardson developed Formula KR20 and KR21 to estimate the internal consistency reliability by using Method of rational equivalence. KR21 formula is the simplified version of KR20 with slightly lower accuracy. KR21 is useful when all the questions in the test have same level of difficulty.

$$\text{KR 20}: \rho_{KR20} = \frac{K}{K-1}\left(1 - \frac{\sum_{j}^{k\,j=1k} p_j q_j}{\sigma^2}\right)$$

$$\text{KR21}: \rho_{KR21} = \frac{K}{K-1}\left(1 - \frac{\mu(K-\mu)}{K\sigma^2}\right) \text{ Where,}$$

$P_j q_j$ = Product of proportion of respondents with right answer (*p*) and proportion of respondents with wrong answer (*q*)

K = Number of items within test

σ^2 is the population variance of total score obtained by each individual μ is the mean total score

Method of rational equivalence is superior over split half method because split-half method simply emphasizes on the equivalence whereas rational equivalence method emphasizes on both equivalence and homogeneity. Additionally it neither requires test to be split nor requires double administration of the test.

Factors Affecting Reliability

Factors intrinsic to the test

a. *Length of the test* : As the length of test increases the number of items in the test also increases. With this increases the reliability coefficient also increases. It is similar to the sampling principle that as the sample size increases the accuracy of measure increases and error decreases.

b. *Homogeneity of items* : When the test is measuring unidimensional phenomenon –as the number of homogenous test items increase the reliability of measuring instrument also increases.

c. *Difficulty level of items* : Too difficult and too easy tests result in lower reliability due to lower spread of test scores.

d. *Test Instruction* : Clear and concise instructions to the respondents give valid estimate of reliability whereas ambiguous and lengthy instructions result in lower reliability.

e. *Item selection* : If there are too many interdependent items in a test, the reliability is tends to be lower.

Extrinsic factors

a. Mood of the respondents

b. Heterogeneity in the respondent group

c. Grasping power and level of understanding on the side of respondent

d. Physical and social environment existing while administering the test

e. Method of administration.

How to Improve the Reliability

a. If reliability is low, add more homogenous items

b. Use unambiguous, simple and concise words and sentences while writing the test items

c. Give clear instructions to the respondents

d. Administer test/scale under standard, well controlled and similar conditions.

Practical Importance of reliability

One the major instrumental purpose of scientific enquiry is to discover the underlying relationship between variables. If the data gathered from a measuring instrument is unreliable then the relationship discovered by using such data are dubious and predictability of the phenomenon becomes difficult by using such data. Even though reliability is not the most important facet of scientific research, there can be no good scientific work without reliable data. To summarize, reliability is a necessary but not sufficient condition for the scientific research.

Case studies

Case 1: An extension researcher –in order to test knowledge level of the farmers about the Mango production practices –has developed a knowledge test. Prior to using of the test for measuring the knowledge, he is interested in estimating the reliability of the results generated by the knowledge test. In order to do that, researcher used the split half method to estimate the internal consistency by splitting the 20 questions in to two halves. Following table depicts the score obtained by the same group of respondent farmers for the two subsets of the test. Estimate the internal consistency reliability by using split half method.

Respondent No.	1	2	3	4	5	6	7	8	9	10	11	12
Even Question subset	22	18	23	21	14	15	21	17	17	16	19	23
Odd Question subset	20	15	21	24	16	11	24	18	23	16	15	22

Solution

First calculate the Pearson correlation coefficient for the given dataset. (This can be done by using "CORREL" function in MS Excel or use conventional Pearson's correlation coefficient formula to calculate manually).

Here in this case Correlation Co-efficient (r) = 0.6672

Now use the Spearman-Brown formula

Where,

r = Inter-correlation between two sets of scores

So reliability of the knowledge test here is 0.80.

Case 2: A researcher developed an eleven questioned dichotomous test for assessing the farmers' awareness about utility of mobile phones in seeking agricultural information. He is interested in knowing if the items in the awareness test are homogenous or not. To do this, he administers the test to a group of 12 farmers. Table given in page No. depicts the response pattern of the farmers. Estimate the internal consistency reliability of the test items by using the Kuder and Richardson Formula (Assume that 1 = Right Answer and 0 = Wrong Answer).

Steps

- Count the number of questions (K) = here 11 in this case
- Calculate the total score obtained by each respondent farmer (sum of Q1 to Q11 for each of the respondent farmer)
- Calculate the Proportion of respondents with right answer for each f the question. (Example in case of question number 7, 5 respondents have given right answer. That means the proportion of respondents with right answers in this case is 0.4167)
- Similarly calculate the Proportion of Respondents with wrong answer (q) for each of the question. Alternatively you can get q value for each of the question by using the formula $q = 1 - p$

  Now for each of the question find the product of p and q. (for Example in case of Question No. 7 p = 0.4167, and q= 0.5833, so pq = 0.2431
- Now add all these products obtained in previous step to get sum of pq (depicted as) Σpq
- Subsequently calculate the score variance (σ^2) for the total scores obtained by the all the respondents. Here data points for calculating the variance are given in last column. (Note: here we are supposed to calculate population variance and not the sample variance, so we will use the population variance formula rather than sample variance formula. While calculating variance in MS Excel, be sure to use VARP command rather than VAR command)
- Now use the formula $r_{KR20} = \frac{K}{K-1}(1 - \frac{\Sigma pq}{\sigma^2})$ to estimate reliability of the test.
- Interpret the reliability coefficient.

Farmer	Q1	Q2	Q3	Q4	Q5	Q6	Q7	Q8	Q9	Q10	Q11	Tctal Score
1.	1	1	1	1	1	1	1	1	1	1	1	11
2.	1	1	1	1	1	1	1	1	0	1	0	9
3.	1	0	1	1	1	1	1	1	1	0	0	8
4.	1	1	1	0	1	1	0	1	1	0	0	7
5.	1	1	1	1	1	0	0	0	1	0	0	6
6.	0	1	1	0	1	1	1	1	0	0	0	6
7.	1	1	1	1	0	0	1	0	0	0	0	5
8.	1	1	1	1	1	0	0	0	0	0	0	5
9.	0	1	0	1	1	0	0	0	0	1	0	4
10.	1	0	0	1	0	1	0	0	0	0	0	3
11.	1	1	1	0	0	0	0	0	0	0	0	3
12.	1	0	0	1	0	0	0	0	0	0	0	2
Total	**10**	**9**	**9**	**9**	**8**	**6**	**5**	**5**	**4**	**3**	**1**	**69**
Proportion of Respondents with correct answer (p)	0.8333	0.7500	0.7500	0.7500	0.6667	0.5000	0.4167	0.4167	0.3333	0.2500	0.0833	
Proportion of Respondents with wrong answer (q)	0.1667	0.2500	0.2500	0.2500	0.3333	0.5000	0.5833	0.5833	0.6667	0.7500	0.9167	
pq	0.1389	0.1875	0.1875	0.1875	0.2222	0.2500	0.2431	0.2431	0.2222	0.1875	0.0764	Σpq = 2 **1458**
K	11											
Score variance ($\acute{o}^2$)	6.5208											
r_{KR20}	0.7380											

Exercise

State whether below given statements are true or false and justify your answer

1. While assessing the internal consistency reliability, if the test is reliable, then each possible subset will yield approximately the same rank order for the individuals.
2. Reliability is an absolute measure.
3. Reliability is aimed at answering question "how accurately we are measuring what we are supposed to measure?"
4. Higher the value of split half reliability coefficient, more heterogeneous the test items are.
5. Reliability can be measured without knowing what the instrument was supposed to measure.

Chapter 9

Validity of the Measuring Instrument

Reliability aims at assessing the accuracy and consistency of whatsoever the measuring instrument measures irrespective of whether we are measuring what we are supposed to measure or not. Whereas the validity emphasises on the question *"are we measuring what we were supposed to measure?"*

To clarify the difference between reliability and validity, let us take an example of instruments that measure colour intensity. Assume, an expert develops two separate instruments X and Y to measure the intensity of green and red colour, respectively. But, the commercial manufacturer wrongly tags the red colour measuring instrument as X. When someone uses this wrongly tagged instrument for measuring the green colour intensity, whatever intensity values the instrument yields are reliable but not valid. It is so because the instrument is measuring red colour intensity *accurately* (hence data is reliable) but it is not measuring what it was supposed to measure –the green colour intensity (hence data is not valid).

Above said example makes the difference between reliability and validity quite clear. Reliability deals with precision and consistency of measurement whereas the validity is related more with the construct to be measured. It is possible to study the reliability even without knowing what variable/was supposed to measure, whereas it is not possible to study the validity without knowing what variable/ was supposed to be measured. *Validity can be defined as the extent to which a measuring instrument measures what it was supposed to measure.*

Types of Validity

a. Content Validity *(Not to be confused with face validity)*

b. Criterion related Validity

c. Construct Validity

Content Validity

Content validity also known as logical validity is the representativeness or the sampling adequacy or the measuring instrument. It asks the question *"to what extent the content of the test/scale is representative of the domain of the psychological object under investigation*?" For example, when a trainer is interested in measuring the effect of training on the knowledge level of the farmer he will initially construct a knowledge test and subject it to content validity in order to determine to what extent the questions/items in the knowledge test are representative of the training course content.

Whether content validity is same as face validity?

Sometimes content validity is interchangeably used with face validity. But in fact both are different. *Content validity should not be confused with face validity*. Technically speaking, face validity is not actually validity. It refers not to what actually test measures, but to what it appears superficially to measure. Face validity pertains to whether the test looks valid to the examinees who take it, the administrative person who makes decision about using the measuring instrument and other non-technical observers who are not concerned with technical aspect of test / scale construction. If the test is known to have content validity, face validity can be assumed. However, it does not work in reverse direction. Face validity does not ensure content validity.

Content validity is more efficient when the construct under investigation is operationally defined indicating entire range of observable behaviours and natural phenomenon that come under the psychological domain of the construct. It requires the judgement of the subject matter specialist who has experience in investigating the psychological object of the concerned study. Face validity is concerned about two questions—one, whether the given item in the test/scale measure what it is supposed to measure and secondly to what extent the entire range of items given in the test/scale are capable of measuring the entire domain of the psychological object under investigation (i.e. it is concerned with to what extent the items in the test/scale are representative of the psychological domain of construct under investigation).

Preconditions for Content validation

a. Each item must be judged separately for its presumed relevance to the property being measured.

b. Judgement should be done by the experts who have experience in studying /observing/experiencing the phenomenon under investigation.

c. It is not sufficient if all the items are measuring what they were supposed to measure.

d. It should be ensured that the range items are capable of catching every dimension of the construct under investigation.

Content validity is judgemental validity and it is not a comprehensive measure of validity. Hence in most of cases content validity alone cannot be used as a measure of validity. It must be combined with other methods of validation for better results.

Criterion related validity

In case of criterion related validation method, even though there are two different approaches –*concurrent and predictive* – the primary interest of the investigator is the criterion which he wishes to predict. In both the cases he administers the test, obtains an independent criterion measure on the same subjects and compute correlation. If the test score and criterion (existing well established measuring procedure) score are obtained at the same time then it is *concurrent validity*. If the test is administered and scores are collected earlier and the criterion measures are studied later at some point of time in future then it is *predictive validity*. The major requirement of the criterion related validity is that the criterion and new measurement procedure should be theoretically related.

Even if well-established standard measuring procedures exist, sometimes we may need to create a new measuring instrument either to create its shorter version or to account for the new context, location, culture etc. or to help test the theoretical relatedness of the criterion with the construct it is measuring. Concurrent validity is used generally when a test is constructed to replace an existing test that measures the same construct. Predictive validity is generally used to demonstrate that given measuring instrument makes accurate predictions about the construct it is supposed to measure. In case of predictive validity one does not care about what is being measured by instrument but one cares for its predictive ability. It is much concerned about the practical problem and outcomes.

An example of concurrent validity: Suppose we are interested in developing a new knowledge test for measuring the knowledge level of farmers regarding adoption of plant protection measures in case of rice. We already have a test but it cannot be used to measure knowledge level because that was developed for some other region. So in this case a sample group of farmers will complete the two tests simultaneously –newly developed test and the previously developed standard test. We want to know whether the new measurement procedure really measures farmers' knowledge level regarding plant protection measures in rice. If it does then it must demonstrate strong and consistent relationship between the scores from the new measurement procedure and the scores from the well-established measurement procedure. This is often measured using a correlation efficient. If the relationship is inconsistent or weak, the new measurement procedure does not demonstrate concurrent validity.

Suppose we have developed a test to select the students for admission in the college based on their intellectual ability, so that we should be able to select those students who are most likely to give relatively higher academic performance. We now want to check that to what extent the intellectual ability measured by this test is capable of predicting the academic performance. In this case we will correlate the intellectual ability test scores of the students (measured prior to getting admitted) with the GPA (grade point average) of the respective student (measured at the end of the year). If the correlation is strong and consistent then it means that test has high predictive validity. If the correlation is weak and inconsistent then it means the test does not have acceptable level of predictive validity.

Construct Validity

It is the most advanced format of validity because it unites psychometric notions with theoretical notions. Construct validity emphasises on the question "*what constructs account for variance in test performance*?" Here investigator seeks to explain individual differences in the test scores of the measuring instrument. The point of interest in this case is the property being measured rather than test itself.

Construct validity refers to the basis of the causal relationship and is concerned with the congruence between the study's results and the theoretical underpinnings guiding the research. In essence, construct validity asks the question of whether the theory supported by the findings provides the best available explanation of the results (Marczyk, DeMatteo and Festinger, 2005).

Strategies for Improving Construct Validity (Cook and Campbell, 1979)

i. Provide a clear operational definition of the abstract concept or independent variable.

ii. Collect data to demonstrate that the empirical representation of the independent variable produces the expected outcome.

iii. Collect data to show that the empirical representation of the independent variable does not vary with measures of related but different conceptual variables.

iv. Conduct manipulation checks of the independent variable.

Summary of Major Types of Validity of a Measuring instrument

Content validity	:	The content of the test appears to experts to accurately represent the full range of what the test claims to measure.
Criterion-related Concurrent validity	:	Scores on the given test are correlated with some other measures of the same construct which test is supposed to measure.
Criterion-related Predictive validity	:	The test scores used to predict scores on another variable that is supposed to predict the variable of interest; these scores are correlated with the observed measures of the variable under investigation in near or far future.
Construct validity	:	There is evidence that what the test measures corresponds to the underlying theoretical variable it is meant to measure/assess (Aron, Coups, and Aron 2012).

Exercise

1. For a measuring instrument which is more important, Reliability or Validity? Justify your answer.
2. Differentiate between reliability and validity on at least three aspects.

References

Aron, A., & Coups, E. (2013). *Statistics for Psychology* (6th ed.). Boston: Pearson.

Cook, T.D. and Campbell, D.T. (1979). *Quasi-Experimentation: Design and Analysis for Field Settings*. Rand McNally, Chicago, Illinois.

Chapter 10

Internal Validity

In previous sections we discussed about the validity and estimation of validity. Now the question is whether the reliable and valid measuring instrument is sufficient condition to ensure the reliable and valid results in actual experiments? What if the measuring instruments are perfectly reliable and valid but the experiment is conducted in less controlled conditions? Can we expect accurate predictions under such conditions? The answer is No. This is so because, even though the measuring instrument measures accurately and consistently what it was supposed to measure, the variable/phenomenon *per se* is influenced by the unwanted and unintended variables. So measured changes are accurate but the changes in the phenomenon/event/variable being measured is influenced by extreneous variables, so relation established by using less controlled experimental procedures is less accurate and sometimes fails to trace down the weaker relationships and the confidence in claiming such relationships is very low. On other hand the relationships established among two variables or predictions made via the experiments conducted in less controlled conditions are more similar to real life situations and hence such predictions/relations are more practical and observable in natural settings and hence can be easily generalised.

The above given discussion about the accuracy of establishing relationship and the generalizability of the research findings gives rise to the concept of internal and external validity. Note that internal or external validity must not be confused with validity of measuring instrument. Internal and external validity are broader aspects and are defined in relation to entire research design, whereas previously discussed validity was defined specifically in relation to the measuring instrument.

Internal validity: Although evidence of absolute causation are rarely achieved, the goal of most experimental designs is to demonstrate that the independent variable was directly responsible for the effect on the dependent variable and ultimately the results found in the study. In other words the researcher ultimately wants to know whether the observed effect or phenomenon is due to the manipulated independent variable or due to some uncontrolled or unknown extraneous variable or variables (Pedhazur & Schmelkin, 1991).

Internal validity refers to the ability of a research design to rule out or to make implausible alternative explanations of the results or plausible rival hypotheses (Campbell, 1957; Kazdin, 2003). A *plausible rival hypothesis* is an alternative interpretation of the researcher's hypothesis about the interaction of the independent and dependent variables that provides a reasonable explanation of the findings other than the researcher's original hypothesis (Rosnow & Rosenthal, 2002).

In simple words, internal validity may be defined as the ability of research design to conclude that the experimental (independent) variable did indeed create change in the dependent variable. *It is the ability of the research design to accurately spell out the relationship between cause and effect variables, excluding the effect of extraneous and confounding variables.*

Practical importance of internal validity in extension impact assessment: Internal validity is of great importance especially for the socio-economic impact assessment studies. When we are assessing the socio-economic impact, our aim is to assess that to what extent changes observed in community (*effect*) are dues to the programme interventions (*cause*). Internal validity in this kind of studies is the ability of research design to exclude the effect of alternate possible causes (plausible rival hypotheses) of the observed change in dependent variable (impact) and detect the portion ofvariation in dependent variable (impact) due to programme interventions. It is the ability of research design to accurately assess the impact on programme beneficiaries as a result of programme interventions excluding the changes that are brought by other programmes and factors operating simultaneously. If the internal validity is not sufficient then it is difficult to attribute the social impact to a particular programme. Programme impact assessment studies with very low internal validity are very harmful for decision and policy makers.

Threats to the internal validity: There are many factors that reduce the internal validity of the research design. Below discussed are the few of the most frequently observed threats to the internal validity.

***History*:** Human behaviour is dynamic and it responds to the internal or external phenomenon and incidents that take place around him/her. Every research has a time dimension. When time dimension is involved, there is possibility of occurrence of certain events between two time points. Likewise when our research process involves time and human dimension the respondents are bound to be exposed to unintended and uncontrolled incidents which will bring about change in knowledge, skill, experience and insight of the respondents. These changes inturn will affect the studies final outcome or create confounding effect and reduce the internal validity of the study.

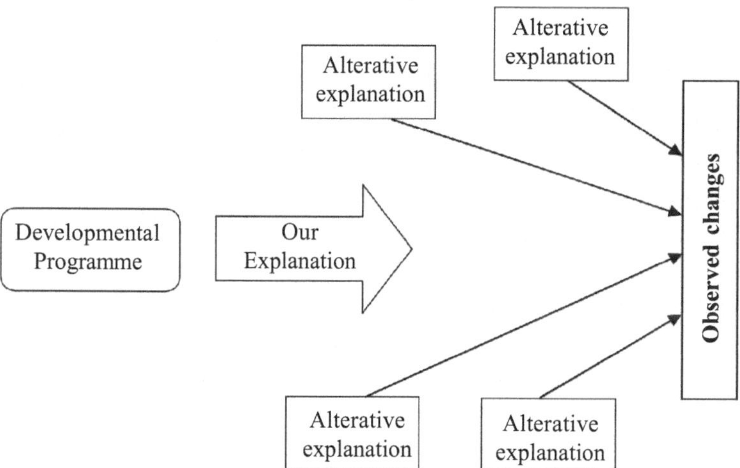

Maturation: Time dimension also yields another negative impact on internal validity of the research design. With time individuals mature and are bound to go through intrinsic changes which in turn will reduce internal validity of the design. *Example:* When we are studying the impact of long term training programme on children's mathematical ability the progressive change in their mathematical ability are combined effect of the training programme and that of their mental maturity.

Instrumentation: It refers to changes in the recorded observations of independent variable due to change in the measuring instrument, measurement procedures or measuring personnel over a period of time. When we are doing long term impact studies, when observers changes and the measurements are more quantitative and judgemental in nature then we will observe high variations in measurement.

Testing: It refers to effect of pretesting on the performance in post-tests. Pretesting generally increases confidence of the respondent, creates boredom, negligence and memory effect which inturn affect the post-test performance and reduce internal validity of the study. If we administer the same reading ability test before and after a training programme generally we will see more fluent reading in the post test. This holds true even for control group which has not received any training.

***Selectionbias*:** Thumb rule in sampling and assignment of treatment is *"select the respondents randomly from the group, randomly assign individuals to the group and randomly assign groups to the treatment"*. If this rule is not followed selection is biased and will affect the internal validity of the research design.

***Experimental Mortality*:** This means loss of subjects in the study over a period of time. This in most common in long term experiments where we expect respondents to be part of experiment –may be for many months or years.

***Regression effect / regression to the mean*:** It is a statistical phenomenon and occurs when we proceed with a non-random sample from a population and two measures that are imperfectly correlated. This phenomenon is highly observable when sample includes one of the extremes of the normally distributed population. For example when we administer a test to the group of respondents and again administer the test to those respondents of the same group who scored lowest in the first testing –irrespective of any treatment in between two testings –then obviously lowest scoring portion of the group automatically shows some increment in the average score. This is so because, practically the lowest scoring group was at its worst performance and it cannot go below that. *"You can only go up from here"* phenomenon operates in this case.

For example, if we select a sample of only the students who constitute the lowest scoring ten percent of the class on the pre-test, then it is very unlikely that the same sample once again constitutes exactly the lowest ten percent scoring portion on the post-test. Most of them would score low on the post-test, but they aren't likely to be the lowest ten percent twice. For instance, maybe there were a few students on the pre-test who luckily were right on a few guesses and scored enough to reach at the eleventh percentile who won't get so lucky next time. If you choose the lowest ten percent on the pre-test, they can't get any lower than being the lowest –they can only go up from there.

Note that these threats are inherent in single group test research designs. To eliminate one or more of these threats, we usually consider going for two or multiple group research designs including one of them receiving no treatment – called as control group. A control group is one which receives no experimental treatment. Using of two or more group in research design reduces the effect of above discussed threats of internal validity but do not completely eliminate these threats. We will discuss in consequent chapters about the research designs and their types along with merits and demerits of each of the design.

Social Interaction Threats

These are not absolutely different class of threats but fall under one or more categories of threat discussed above, but these types of threats are results of the interaction among the individuals in the social system.

Diffusion and Lmitation Threat: This type of threat exists when a control/ another treatment group learns about the unintended treatment either directly or indirectly from some other treatment group. This kind of threat tends to nullify the difference between control and treatment group.

Compensatory Rivalry: Here the control group knows about what treatment group is getting and tries to perform better than usual and tries to outperform the treatment group, so this kind of threat also tends to nullify the difference between control and treatment group.

Resentful Demoralization: Here the behaviour of the control group is opposite to that of control group's behaviour in previous case. Here the control group knows about what treatment group is getting and feels discouraged and gives up leading to poor performance than usual. This kind of threat tends to amplify the effect of treatment and sometimes may establish the relationship which doesn't exist truly.

How do we deal with threats to internal validity?

There are several alternatives to eliminate/ minimize these threats; one of the most commonly used approaches is the moderating of research design. As discussed earlier one of the best moderation that can be done in research design is usage of the one or more control groups and using of various treatment groups. Using of control group to certain extent reduces the threats of maturation, history, testing effect, instrumentation issues, experimental mortality and regression to the mean. But it introduces the social interaction based threats like diffusion and imitation threat, compensatory rivalry and resentful demoralization. These social interaction based threats are easy to control as compared to other threats of internal validity and hence the inclusion of control group in research design is practically more effective in improving internal validity.

Another best way to improve the internal validity is to conduct the experiment in controlled conditions. Lower the interference of extraneous variables higher the chances of having greater internal validity for the research results.

There are some complex research designs like Solomon four group design which utilize more than one control groups in order to make research design more efficient by minimizing the influence of threats of internal validity.

Other important methods of controlling the effect of threats to the internal validity include random selection of subjects, random assignment of subjects to the groups and random assignment of the treatment to the group and use of statistical methods. Randomization is a control method that assists in ensuring that extraneous sources of bias and artifact will not confound the validity of the findings of the experiment. Randomization helps to ensure the internal validity of the study by eliminating alternative rival hypotheses that might explain the results of the study

Exercise

1. Select three research papers from your area of interest and identify the possible threats to the internal validity of the findings
2. Please denote whether below given statements are true or false, along with your own justification
   a. Internal validity is more concerned with accuracy in establishment of causal relationship
   b. Internal validity is more concerned with predictability of cause effect relationship in natural setup
   c. Testing effects exists only in experimental designs that involve pre-test and post-test
   d. Instrumentation threat exists only in experimental designs that involve pre-test and post-test
   e. Probability of regression effect is high when we asymmetrically sample the population.

_ _

_ _

_ _

_ _

_ _

_ _

_ _

_ _

_ _

References

Campbell, D.T. (1957). *Factors Relevant to the Validity of Experiments in Social Settings.* Psychological Bulletin, 54, 297–312.

Kazdin, A. E. (1992). *Research Design in Clinical Psychology* (2nd ed.). Boston: Allyn& Bacon.

Marczyk, G., and DeMatteo, D. (2005). *Essentials of Research Design and Methodology.* Hoboken, N.J.: John Wiley & Sons.

Pedhazur, E. and Schmelkin, L. (1991). *Measurement, Design, and Analysis: An Integrated Approach.* Hillsdale, N.J.: Lawrence Erlbaum Associates.

Rosnow, R. L. and Rosenthal, R. (2002). *Beginning Behavioral Research: A Conceptual Primer.* (4th ed.). Upper Saddle River, NJ: Prentice Hall.

Chapter 11

External Validity

In previous section we discussed about the ability of the research design to precisely detect the relationship between cause and effect under controlled conditions. But in social research it becomes difficult to conduct the research under highly controlled conditions and moreover many of the researches in social science aim at prediction rather than understanding.

As we discussed earlier, aim of science is development theory, understanding, prediction and explanation of natural phenomenon. In many of the cases we are more interested in predicting the occurrence of social phenomenon rather than explaining it or understanding its relationship with other phenomenon. Under such conditions internal validity is of less use because a relationship that is established in artificially controlled conditions (with high internal validity) may fail to express itself in natural setup due to effect of extraneous variables. When the relationship between cause and effect is weak enough to be expressed in natural setup then it is of less use in predicting the phenomenon. Under such conditions we require experiments that are conducted in nearly natural setup, so that we can generalize the findings of the research to the wider domain. So what exact external validity means?

External validity refers to the degree to which the research results can be generalized to other conditions, participants, times, and places. External validity refers to conclusions that can be drawn about the strength of the inferred causal relationship between the independent and dependent variables to circumstances beyond those experimentally studied.

Note that in all forms of research design results and conclusions of the study are limited to the participants and conditions as defined by the contours of the research. External validity of a research design refers to the extent to which the findings of an experiment can be extended or generalized to the entire population/natural setup/other similar conditions. *External validity refers to the generalizability of the results of a research study.*

Threats to External Validity

External validity is more important from the prediction point of view in natural set up. The external validity of research is affected negatively by many of the factors like sample characteristics, stimulus characteristics and settings, reactivity of experimental arrangements, multiple-treatment interference, novelty effects, reactivity of assessment, test sensitization, and timing of measurement (Kazdin, 2003). Developing research design considering the influence of these factors will be helpful in confidently generalizing the results of the study to other circumstances and populations. Below listed are the most commonly faced threats to the external validity.

Sample Characteristics

It is the situation where the sample is relatively peculiar and hence the results of the design are applicable only to a particular sample. Under such conditions it will be unclear whether the results of the study can be generalized to other samples or not.

Example: A researcher found that certain psychological training programme is effective in significantly enhancing the achievement motivation of the members of mango growers association. Whether the results obtained based on farmers association can be extended to the Women SHG members? Can we use the results of previous study to interpret that the given psychological training programme will significantly enhance the achievement motivation of the members of Women SHG?

Generally farmers' producers association are dominated by male farmers, whereas the women SHGs members are females. So both these group will defer with respect to sample character –gender. Hence the generalizability of the previous study to the women SHGs remains doubtful.

Sample characteristics may include a wide range of traits and demographic characteristics. Some of the most commonly prevalent sample characteristics are age, gender, education, and socioeconomic status. Some diversity-related characteristics are not well represented in most forms of research. The primary area of concern here is the overrepresentation of some sections of sample, such as male respondents; and a related, limited inclusion of underrepresented and minority groups, such as female respondents. Sample diversity characteristics are important factors in the context of external validity and they can have far-reaching serious consequences for all strata of society.

Stimulus Characteristics and Settings

It refers to an environmental phenomenon in which peculiarity of the study conditions limit the generalizability of the findings to the other conditions.

Every study operates under a unique set of conditions and circumstances related to the experimental arrangement. The most commonly cited examples include the research setting and the researchers involved in the study. The major concern with this threat to external validity is that the findings from one study are influenced by a set of unique conditions and thus may not necessarily generalize to another study, even if the other study uses a similar sample.

Reactivity of the Experimental Arrangements (Hawthorne effect)

Reactivity of the Experimental arrangements also called as hawthorne effect refers to a potentially confounding variable that is a result of the influence produced by knowing that one is participating in a research study (Christensen, 1988). Participants' awareness about being part of a research study may increase the participants' attentiveness; some participants might be more compliant diligent while others might intentionally show difficult or noncooperative behaviour. This affects participants' attitudes, behaviour and their performance during the course of the study. This in turn can have a significant confounding impact on any results obtained from the study and is especially more serious when participants are aware about the purpose of the study. In the context of external validity, the question we ask here is whether the same results would have been obtained provided the participants were unaware about their participation in any such study (Kazdin, 2003). The primary source of this threat is the ethical standards requirement that participants' informed consent is necessary for including them in research study.

For example, let's consider a study designed to assess the effectiveness of a behaviour, modification training designed to reduce test anxiety of students. The experimental group undergoes the training programme and the control group does not. The researchers found that the experimental group shows lower levels of test anxiety as compared to the control group –post-treatment. Based on this result, the researchers might be convinced to say that the intervention was responsible for the reduction in test anxiety; however, it might be that the behaviour under investigation improved because the participants had assumed a compliant attitude toward the intervention. On the other hand, if the participants in the treatment group had adopted a pessimistic attitude towards the training programme, then the results of the study might have implicated that the training programme was not effective in reducing test anxiety. In any event, either outcome might be the result of reactivity to the experimental arrangements and not the intervention itself.

Multiple-Treatment Interference

This refers to research circumstances in which (1) participants are administered multiple experimental treatments within the single study or (2) the same individuals participate in more than one study (Pedhazur&Schmelkin, 1991). Even though it is most common in cause-effect studies, it is also common in any research study that is characterised by multiple experimental conditions or independent variables. The major implication of this threat is that the research results may be due to the context or series of conditions in which the research is presented. (Kazdin, 2003).

Novelty effects: This refers to the possibility that the effects of the independent variable may be due in part to the uniqueness or novelty of the stimulus or situation and not to the intervention itself. It is similar to the Hawthorne effect in that the new or unusual treatments or experimental interventions might produce results that disappear once the novelty of the situation or condition wears off. In other words, the novelty of the intervention or situation acts as a confounding variable and it is that novelty (and not the independent variable) that is the real explanation for the results.

For example consider a situation in which researchers are trying to determine the effectiveness of a new media lab oriented training programme for enhancing farmers learning of the subject matter. Initially farmers' rate of learning is assessed and then the farmers are randomly assigned either to a control group or one of three treatment conditions. The three learning conditions administered were Lab training, Field training and Classroom training. All of the participants were subjected their respective treatments for 4 weeks and are then revaluated for their rate of learning. The researchers found that Lab training is more effective than Field training and Classroom training.

Here researcher is tempted to conclude that the Lab training is more effective than the Field training. But the reason may be that the Lab training is more novel method for farmers and it creates lot of curiosity to learn as compared to field method –which is a part of farmers day to day life and have no novelty in it. As farmers get accustomed to the Laboratory method of training the curiosity will be reduced and we may further find that Field method is in fact more effective in long run.

Reactivity of assessment: This is quite similar to the Hawthorne effect but not same. In case of Hawthorne effect the participants were aware about their part of being experiment and hence may show positive or negative behaviour based on their attitude towards experimental purpose rather than the set up.Whereas in case of Reactivity of assessment subjects are aware that their performance is being measured and hence they will perform relatively better than they do normally. So reactivity of assessment is the phenomenon in which participants' awareness that their performance being measured can alter their performance from what it otherwise would have been.

For example an educational psychologist want to know the effectiveness of certain reward oriented teaching method over the punishment oriented teaching method. Researcher selected Class IX students from two Schools R (reward oriented teaching is followed) and P (Punishment oriented teaching is followed). Students in both the schools were informed about the purpose of the study. But at the end of experiment the researcher found that the academic performance was relatively better in School P –where punishment oriented teaching method was followed.

Researcher in this case may be tempted to say that the punishment oriented teaching is more effective than reward oriented teaching. But in reality the case may be that, the students in School P might have took it as a matter of pride & competition with the students of School R which motivated them to outperform irrespective of the teaching method. So the students' awareness –about the fact that their performance is being measured –acted as confounding variable in this case and is a threat to the external validity of the research results.

Pretest and posttest sensitization: These threats refer to the effects that pretesting and posttesting might have on the behaviour and responses of the study participants. Here the question in the context of external validity would be that, would the results of the study have been the same if the pretest had not been administered? Would the same results have been found if the posttest had not been administered?

For example, prior to the assessing the effectiveness of an anxiety therapy students were administered with a pretest to determine their initial anxiety level. Then the students were treated with the anxiety therapy for a time of 7 days. Later on same group of students were administered a post-test by using same anxiety test which was used in pretesting. The result of the posttest showed that students exhibited relatively lower anxiety levels. Even though it seems that therapy has reduced the anxiety level effectively, but in absence of the pretesting, this might not have been the case. Pretesting sensitized the subjects and led to higher scores in posttest.

As a threat to external validity, the concern is that exposure to the pretest may contribute to, or be the sole cause of, the observed changes in the dependent variable. This has obvious implications for external validity because pretest sensitization might render the results irrelevant in situations in which the same pretest was not administered.

Timing of assessment and measurement: This threat refers to whether the same results would have been obtained if measurement had occurred at a different point in time (Kazdin, 2003). This is particularly common in longitudinal research studies.

Longitudinal research is characterized by multiple measurements all along the time span of research process. For example, a longitudinal study on training outcome may reveal that a given training method is relatively more effective in enhancing content learning among the farmers as compared to conventional method. But on long run the results may show that the particular training method is not as effective as the conventional one. The more specific conclusion is supported by the study while the more general conclusion about effectiveness might not be accurate due to the time point of measurement. The reverse also might be true. In some case some treatments may fail to show significant impact but in long run the impact may become more evident and clear. Hence the timing of assessment and measurement of results may emerge as a serious threat to the external validity of the research.

How to increase external validity of research result

a. External validity of the research results can be increased by randomization of the experiment. Randomization eliminates the effect of selection bias and ensures that the selected sample is representative of the population as a whole. This increases external validity of research results.

b. The next possible method is to use the larger samples. Larger sample increases the possibility of the inclusion of large range of variability and makes it representative of the population.

c. Include the counter measures to eliminate the effect of pretesting and sensitization (using of four group Solomon Design).

d. Decide optimal timing of impact assessment based on the purpose, utility and nature of the project.

e. Eliminate the reactivity of the experimental arrangements by using control groups which are aware about their part of being experiment.

External validity can best be understood as an interaction between experimental setup, participant attributes and their related characteristics. Confidence in generalization of the findings from study is weighed down when the predictor variable interacts with attributes of the participant or any other aspect of the experimental setup to produce the observed results. Therefore the threats to external validity discussed in this unit are far from exhaustive list of threats to the external validity. Depending on the nature of experimental design and the objective of the research study, each study can create unique set of threats to external validity. If experimental control is not possible, the limitations of the study's findings should be discussed in sufficient detail to clarify the relevance and generalizability of the findings.

Exercise

1. Select three research papers from your area of interest and identify the possible threats to the external validity of the findings
2. Differentiate between internal and external validity with example
3. Write a short note on practical importance of external validity (300 words)

References

Christensen, L. B. (1988). *Experimental Methodology* (4th ed.). Boston: Allyn& Bacon

Kazdin, A. E. (2003). *Research Design in Clinical Psychology* (4th ed.). Boston: Allyn& Bacon.

Marczyk, G., &DeMatteo, D. (2005). *Essentials of Research Design and Methodology*. Hoboken, N.J.: John Wiley & Sons.

Pedhazur, E. and Schmelkin, L. (1991). *Measurement, Design, and Analysis: An Integrated Approach*. Hillsdale, N.J.: Lawrence Erlbaum Associates.

Chapter 12

Sampling

Sampling is the process of drawing a part of population supposed to be the representative of that population.

Types: Sampling process can be either ***random** or **non-random***. Most frequently we use random sampling method in our studies. But sometimes the purpose of the study demands for non-random sampling in one or other phase of sampling (*Example*: When we want to study the impact of life skills development programme on the disabled female beneficiaries)

Probability sampling: It is the method of drawing sample by using randomization procedures at one or other stage of sampling.

Non-probability sampling: Type of sampling which do not use randomization procedure for selection of sample.

Random sampling is the process of drawing a proportion (sample) of a population or universe so that all possible samples of fixed size *n* have the same probability of being selected (Kerlinger, 1973).

Probability sampling (Random Sampling)

1. Simple Random Sampling: Here, sample drawing is completely randomized and sample is drawn by using random number tables, chit method etc.

a. *Simple Random Sampling with Replacement:* In this case, after drawing each unit of the sample, it is again considered as a part of population for drawing consequent units of the sample. The probability of selection of each of the subject within a sample remains constant.

b. *Simple Random Sampling without Replacement:* In this case, once the subject is selected then it is not added back to the population from which subsequent units of the sample will be drawn.

Advantages

a. Requires minimum knowledge about the population distribution

b. Free from selection bias

c. Most effective when population is relatively homogenous

d. Easy to calculate sampling error.

Limitations

a. Not a good method of sampling when population is having distinct categories/ heterogeneous

b. This method requires relatively large sample size

c. It also requires entire list of objects in the population to draw the sample

d. This method does not use the available knowledge about population heterogeneity.

2. Systematic Sampling: Here sampling is done by choosing first object randomly and then the subsequent objects are chosen at every n^{th} interval.

For example, a researcher wants to study the socio-economic profile of the customers who visit the branded vegetable store. In order to study the profile of the customers he need to select the sample first. Random sampling at a single time point is not possible since different type of customers visit at different time in a day. Hence, researcher decides to carryout sampling for entire day by selecting first customer randomly, and then selects every fifth customer who exits from the store. This kind of sampling is known as systematic sampling.

Advantages

a. Simple method

b. Sampling is distributed over entire range of population

c. Useful when entire list of objects in the population is not available.

Limitations

a. This method is not truly random.

b. Possibilities of introduction of systematic bias are higher.

3. Stratified Sampling: In this case entire population is divided into various strata such that within strata homogeneity is very high and between strata heterogeneity is very high; subsequently from each stratum pre-determined proportion of sample is drawn.

Types: Based on the proportionality of subsamples we can classify stratified sampling into two types; Proportionate stratified random sample and Disproportionate stratified random sample.

a. *Proportionate stratified random sample*: In this case sub-sample drawn from each of the stratum is proportionate to the relative size of stratum within the population.

b. *Disproportionate stratified random sample*: In this case the drawn sub-sample from each of the stratum may or may not be in proportion to the size of strata.

Example: A researcher is interested in studying the effectiveness of certain training programme on inculcating entrepreneurship among farmers. For this, initially the researcher goes through the list of farmers who were beneficiaries of training programme. Of total 3000 trained farmers 700 were men and 300 were women farmers. Initially the farmers are divided into two strata (men and women farmers). Later on he selects 150 men and 150 women farmers randomly by using random number table. This type of sampling where the population was initially divided into exclusive strata and then equal numbers of respondents are drawn from each stratum irrespective of proportionate size of stratum is known as disproportionate stratified random sampling. Alternatively if the researcher draws the subsample which are proportionate to the relative size of stratum, then it is proportionate sampling. To draw a proportionate stratified random sample, here researcher must select 210 men and 90 women farmers in this case.

Advantages

a. Ensures better representation of each of the stratum in the sample. Thus sample may be assumed to be highly representative of the population.

b. One can draw small sample without losing representativeness.

c. Makes it possible to compare the different strata within population.

d. Highly suitable when population is heterogeneous.

Limitations

a. Requires information related to population heterogeneity.

b. Possibility of faulty classification of strata.

c. This also requires information on entire list of units within population.

4. Cluster Sampling: Cluster sampling is done by dividing entire population into different clusters and then choosing clusters randomly. Consequently, sub-clusters are chosen randomly from each of the cluster and the process is repeated at the lowest level of cluster.

Condition: Units within clusters should be as heterogeneous as possible and two clusters should be as homogenous as possible.

Advantages

a. Less time consuming and low cost.

b. Does not require list of entire objects in population.

c. Highly suitable when population is very large & distributed over wide geographical area.

Limitations

a. Relatively inaccurate: Since cluster is seldom representative of entire population.

b. Assumes cluster to be highly heterogeneous which may not be the case.

Advantages

a. This is a simplest method.

b. Cost effective.

c. If adequate knowledge is there about population, it can result in highly representative sample.

5. **Multistage Random Sampling:** Here the population is divided progressively into smaller and smaller subunits as we move from one stage to another. And from smallest sub-population objects are drawn randomly.

For example, a researcher is interested in assessing the farmers' awareness about blight disease management strategies in paddy crop in 'X' state of India. Initially he identifies the district in with highest area is under Paddy crop. Later the researcher randomly selects two blocks from the selected districts, subsequently he selects two villages and from each village 50 farmers are selected. This type of sampling is called as multistage random sampling.

State – District – Block – Village – 50 farmers drawn randomly

Advantages

a. Sampling stages can be geographically defined

b. Less expensive

c. Requires only sampling frame of smallest unit of sub-population and not that of entire population.

Limitations

a. It partially biases randomization by excluding the larger portion of the population from chance of being selected.

b. For accuracy, it requires relatively larger sample.

Non-probability Sampling

1. **Quota Sampling:** Quota sampling is a non-probability sampling technique wherein the assembled sample has the same proportions of individuals as the entire population with respect to known characteristics, traits or focused phenomenon. In Quota sampling, knowledge of strata of the population-sex, race, region, and so on- is used to select sample members that are representative, typical and suitable for certain research purposes. It derives its name from the practice of assigning quotas or proportions of kinds of people, to interviewers. Such sampling has been used a good deal in public opinion polls.

2. **Purposive Sampling:** Here sample units are selected purposively and not randomly, in an attempt to make the sample more representative of the population by using the experience and knowledge of the researcher. Suppose an anthropologist is exploring the lifestyle of a certain tribal community by using personal interview as a method of data collection. Prior to collecting data, he needs to select respondents. Researcher is interested in exploring the lifestyle of the tribal community. Such study demands for the respondents who have lived within the community for long period of time, participated in almost every social event and are part of community decision making. Under such circumstances, random sampling from the population is not suitable method for exploring the life style. Hence the researcher will purposively select only the experienced, social, old age and moderately aged members of the society based on their insight about the tribal community. This kind of sampling is called as purposive sampling.

3. **Accidental Sampling:** Accidental sampling is a type of non-probability sampling that involves the sample being drawn from that part of the population that is immediately accessible to the researcher. Here sampling is done from whatever in hand available population. Sampling is done at the convenience of the researcher. It is the weakest form of sampling. In effect, one takes available samples at hand. For example, classes of seniors in high school, sophomores in college. This practice is hard to defend. Yet, used with reasonable knowledge and care, it is probably not as bad as it has been said to be. The most sensible advice seems to be: Avoid accidental samples unless one can get no other samples.

4. **Snowball Sampling:** Snowball sampling is a non-probability sampling technique that is used by researchers to identify potential subjects in studies where subjects are hard to locate.

   This kind of sampling is used when the sampling units are sparsely distributed or limited to a very small subgroup of the population. This type of sampling technique works like chain referral. After observing the initial subject, the researcher asks for assistance from the subject to help identify people with a similar trait of interest. The process of snowball sampling is much like asking your subjects to nominate another person with the same trait as your next subject. The researcher then observes the nominated subjects and continues in the same way until the obtaining sufficient number of subjects.

   For example, if researcher is interested in studying the formally less recognized progressive farmers, the researcher may opt to use snowball sampling since it will be difficult to formally less recognized progressive farmers. In this case researcher locates subsequent progressive farmer by using the reference of previously located progressive farmer. Like a snowball increases in size as it rolls down, sample size here increases as researcher proceeds further and further with referrals.

Types of snowball sampling

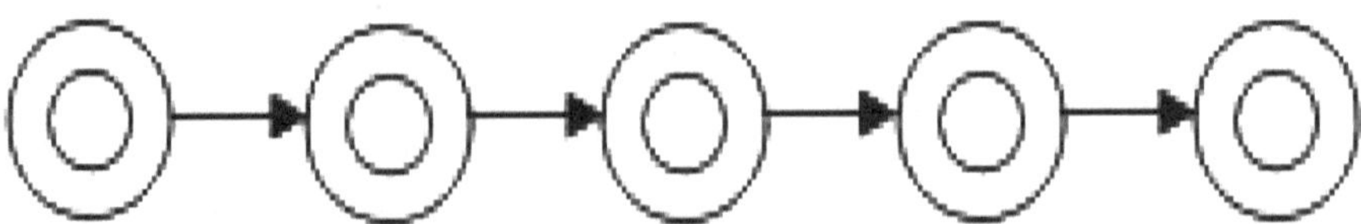

Linear snowball sampling

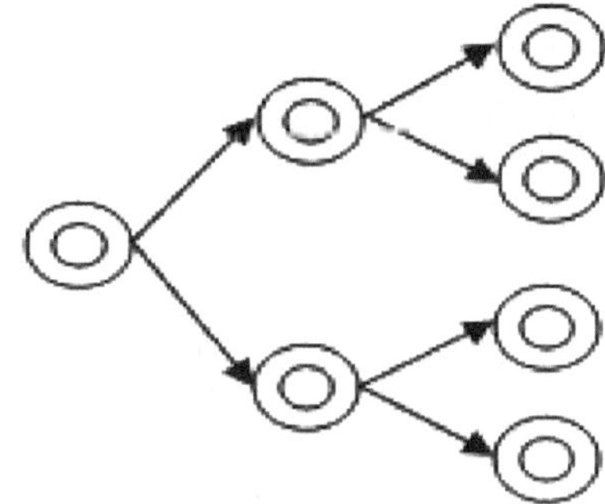

Exponential Non-Discriminative Snowball Sampling

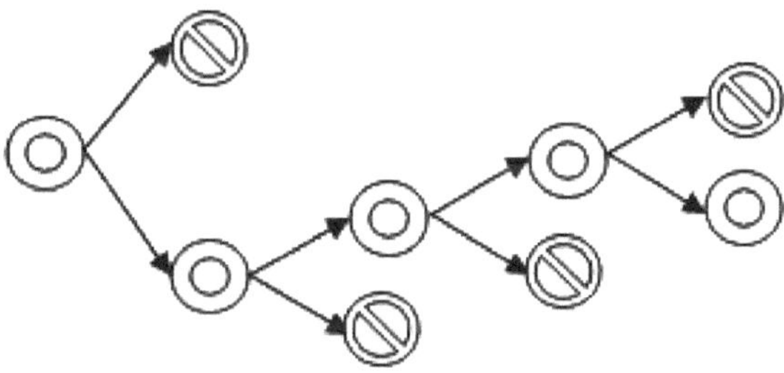

Exponential Discriminative Snowball Sampling

Exercise

1. Differentiate between stratified and multi-stage random sampling.
2. Evaluate and comment on cluster, multistage and stratified sampling procedures used by researchers in real life research studies (use peer reviewed articles to cite real life research examples).

\- -

\- -

\- -

\- -

\- -

References

Kerlinger, F. (1973). *Foundations of Behavioral Research* (2nd ed.). New York: Holt, Rinehart and Winston.

Sukhatme, P. (1954). *Sampling Theory of Survey with Applications*. Bombay: Asia Pub. House.

PART II: RESEARCH METHODS

Chapter 13

Survey Research

Not all the scientific problems demand the detection and establishment of causal relationship among natural phenomenon. Many of the social research problems seek to study the status quo rather than the relation between two or more natural phenomenon. Few of the impact assessment studies (which are kind of causal relationship studies to an extent) also demand for knowing the status quo before and after implementation of interventions. Many other studies demand for benchmark assessment to make policy decisions. Under such circumstances, survey research is a useful research method.

Survey research may be defined as the selecting and studying a population or part of population (sample) in order to discover the coverage, incidence, distribution and intensity of thc socioeconomic and psychological variables among the selected population. Note that we rarely study the entire population (*census*) and most of the time we resort to *sample survey,* because of economic feasibility, time and human resource limitation.

Process of Sample survey

i. Identify the research problem

ii. Define the population

iii. Enlist and operationally define the socio-economic and psychological variables to be observed, measured and recorded

iv. Select suitable test statistics and the level of significance

v. Determine the required sample size (based on required level of confidence in the results and required sample size for the pre-decided statistical test)

vi. Decide upon the sampling method

vii. Develop the reliable and valid measuring instrument

viii. Design and pretest the data collection tools and techniques (questionnaire, interview schedule)

ix. Orient the data collecting personnel to the data collection process

x. Collect the data by using previously developed data collection tools and techniques

xi. Screening and cleaning of data

xii. Process the raw data so that it is amenable to selected statistical analysis

xiii. Analyze the data

xiv. Interpret the results and write final report.

Major types of Social survey methods and their practical suitability

Note these types do not refer to any kind of classification. Types mentioned below are neither exhaustive nor exclusive. Rather, they together form class of social survey methods mostly used or potentially usable in the field of extension education.

Sl. No.	Type of Social survey		When it is suitable
1.	Field survey	:	When we are supposed to interact with the respondents in order to extract required information and when we are interested in exploratory work.
2.	Online survey	:	When we are supposed to collect data from large sample of respondents spread over larger geographical area and the respondents are literate and acquainted in using ICT tools.
3.	Telephonic survey	:	When we are supposed to collect information from population scattered over a larger geographical area and extraction of information from respondents requires two way interaction with the respondents.
4.	Exit poll	:	To assess the voting trend.

Advantages of survey method

a. Enables collection of great deal of information from large sample.

b. Survey research is most suitable for collecting personal and social facts, beliefs and attitude.

c. Survey data can be accurate provided the sampling is probabilistic and sample size is sufficiently large.

d. Access to a variety of participants ensures capturing of possible large variation in population.

e. It is the method where generalized information could be collected from almost any human population.

Limitations of survey research

a. Data collected is superficial.

b. Results have relatively low internal validity.

c. Construct validity of the results is doubtful due to self reporting of the response.

d. Poor sampling leads to inaccurate conclusions.

e. Survey research method is an inappropriate tool for the study of multitudes.

f. Not suitable for studying causal relationships.

Chapter 14

Action Research

Every practical problem need not have a documented solution. Most of the times to solve a practical problem, we need to apply our experience, creativity and experiment with the novel solutions in order to address the problem which have no readymade solution. Under such circumstance, we will apply our knowledge and experience in selecting, planning, experimenting and evaluation of the possible solutions. This type of problem solving approach involves action research. Action research is a type of social research that aims at gaining an insight into the problem by learning from experiences gained in an effort to solve the problem. Here, action itself is a source of knowledge.

Action research is a paradigm and not a method. As a paradigm, action research is a conceptual, social, philosophical, and cultural framework for doing research, which embraces a wide variety of research methodologies and forms of inquiry. Action research is a paradigm that reflects the principle that reality is constructed through individual or collective conceptualizations and definitions of a particular situation requiring a wide spectrum of research methodologies. Characteristically, action research studies a problematic situation in an ongoing systematic and recursive way to take action to change that situation.

Principles of Action Research

a. *Reflective critique:* An individual critically analyses his own experience.

b. *Dialectic critique:* Individual shares his experience to the group and they develop common understanding regarding problem situation and experiences gained in problem solving and learn from each other.

c. *Collaborative resources:* Time, money and knowledge base of the various disciplines is brought together.

d. *Pluralistic structure:* Multidisciplinary and multi-stakeholder based approach.

e. *Calculated risk:* Risk level of failure is rationally selected.

Features of Action Research

a. Problem oriented
b. Source of knowledge is not a theory but a practical action
c. Experiential learning
d. Group sharing and continuous feedback of self-experience
e. Simultaneous solving of problem and generation of new knowledge
f. Co-operative and interactive effort of multiple stakeholders.

Advantages

a. Avoids high theory and complex method
b. Solves problem of immediate returns and generates knowledge as well
c. Understanding about problem.

Limitations

a. Risk of failure
b. Doesn't come out with concrete cause-effect relationship
c. Requires more time and resourceful consumers
d. Does not control extraneous variable.

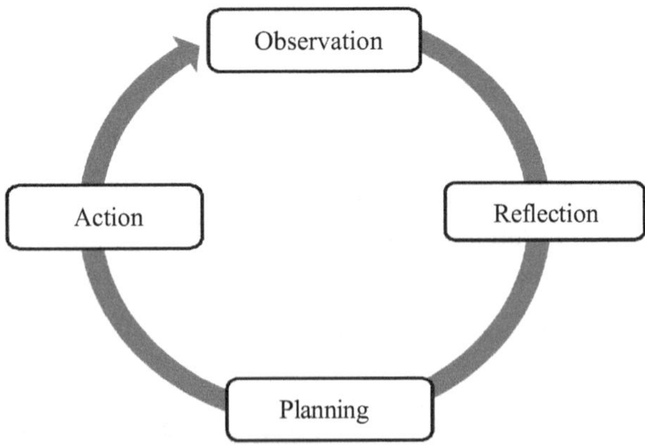

Fig. 1: Process of Action Research

Applications in Agricultural Extension

a. To address novel problems like climate change mitigation

b. To bring about adoption behavior change at community level

c. To control new/ unknown diseases/pests.

Example: Action Research on emerging pest and disease management.

Exercise

1. Identify an action research study in the area of your interest and discuss in detail the procedure of action research used in the study.
2. Discuss in detail how an action research differs from basic research.

Reference

Pine, G. (2009). *Teacher Action Research: Building Knowledge Democracies.* New Delhi: Sage Publications.

Chapter 15

Case Study

Quantitative studies are preferred over qualitative studies for their relatively higher objectivity. But many of the times, complexities involved in the occurrence of phenomenon, its causes, and consequences demand for in-depth analysis of a single unit of study. Whenever we want to study a single unit in depth both qualitatively and quantitatively then case study is the best method of research. Case study is a best tool to use, when we are interested in exploring a new phenomenon on which there is very less scientific information available and also when we are interested in formulating new hypothesis for causal study.

For example when we want to explore socio-psychological factors of entrepreneurial success among small and marginal farmers, case study can be a better choice for in depth study of multiple dimensions rather than a survey research which collects and analyses data at superficial level.

Case study is a non-experimental research technique of exploring and analyzing life of a social unit as a whole, be that is a person, family, community or any other social condition (Kerlinger, 1973).

More complex but more comprehensive definition is that, a case study is *"An empirical inquiry about a contemporary phenomenon (Example, a 'case'), set within its real-world context—especially when the boundaries between phenomenon and context are not clearly evident"* (Yin, 2009).

Characteristics

a. Based on one or very few social unit

b. Conducts in-depth and intensive analysis of social conditions

c. Considers sufficiently wide time cycle

d. Qualitative in nature

e. Does not follow sampling procedure like random sampling

f. It looks at unit of analysis as a whole.

Assumptions

a. Underlying units

b. Influence of time

c. Totality of being

d. Complexity of social phenomenon.

When to use case study method

a. Case study is suitable when your study is trying to provide a descriptive or explanatory answer. *What is happening or has happened? and How or why did something happen?*

b. Secondly, case studies are useful when we want to conduct a longitudinal and in depth study on socio-psychological phenomenon occurring in its natural context.

c. Case studies are suitable when we intend to get more detailed causes and consequences of certain socio-psychological phenomenon, including the emotional aspect of the cause and consequences.

d. When we intend to study the special and rare cases.

Steps in designing and implementation of case study (Based on Yin, 2003)

Define the 'case': The case serves as the main unit of analysis in a case study. At the same time, case studies also can have nested units ("embedded subcases") within the main unit. Put your efforts into selecting and defining as important, interesting, or significant a case as possible. If you are willing to conduct a case study on common everyday phenomenon then, define some compelling theoretical framework for selecting your case. The more compelling the framework, the more your case study can contribute to the research literature.

Select Case Study Designs: Next step is to select one among four types of case study designs. Based on whether your case study consists of single-unit of analysis or multiple units of analysis and whether you are going to keep your case study holistic or you are going to have embedded subcases, there are the possible four types of case study research designs.

- Holistic single case design

- Embedded single case design
- Single case embedded subcase design
- Multiple case embedded subcase design.

Using Theory in Design Work: The third step is related to deciding on whether or not to use theory to support essential methodological steps like developing research question, selecting case, refining case study design, or defining the relevant data to be collected.

Select Data Collection Tools and Techniques and Collect Data: Based on the nature of the phenomenon and the nature of required data, we can collect data using methods like Direct observations, Interviews, Archival records, Documents, Participant-observation and Physical artifacts. There are numerous other sources and techniques for data collection.

Triangulate Evidence from Multiple Sources**:** You should constantly check and recheck the consistency of the findings from different as well as the same sources. In doing so, you will be triangulating—or establishing converging lines of evidence—which will make your findings as robust as possible.

Collect Data on Rival Explanations**:** During data collection, one should scan for events and actions that deviate from our claim and expectations. Data collection should involve a conscious and vigorous effort on search for evidences of discrepancy, as if we are trying to prove the strength of the plausible rival rather than seeking to discredit it. Finding no such evidence, despite an honest and effortful search, increases confidence on the descriptions, explanations and interpretations drawn on the basis of case study.

Data Analysis**:** No single analytical method or set of methods is suitable for analysis of collected data. The context and purpose of the case study demands specific analysis method or combination of various methods of data analysis. Data analysis approaches for case study include but is not limited to Pattern matching, explanation building, time series analysis, replication or corroboratory framework etc.

Presenting Case Study Evidence**:** This is the last stage of case study. Here we communicate the case study findings, support our findings, disprove the plausible rival explanations and point out knowledge gap in order to provide a ground level work for formulation and testing of statistical hypothesis and generation of theory.

Advantages

a. Considers qualitative and emotional aspects of the social unit which the statistical procedure fail to capture.

b. Overcomes the problems associated with detached studies.

c. It is most comprehensive and holistic method as compared to any other method.

d. Provides antecedent variables which researcher may feel to include in laboratory experiments.

Limitations

a. More quantitative

b. Cannot be used for establishing cause-effect relationship

c. Constantly suffers from subject bias of researchers' judgment

d. Costly in terms of time and money

e. Less reliable results and results cannot be generalized to population as such.

Application of case study

a. Provides basic input for formulating the hypothesis

b. Provides understanding about basic domain of social phenomenon for construction of scales/testing

c. Helps in stratification of sample

d. Widens the personal experience of researcher to study deviant behavior.

How case study is different from statistical methods?

Case study	Statistical method
Based on one or few subjects	: Studies large number of subjects
Data collection is intensive and in-depth	: Data collection is extensive and superficial
Selection of units is based on researchers judgment	: By using objective/purposive sampling methods
Consider emotional side of social phenomenon	: Fail to consider emotional side of social phenomenon
Output of research is more narrative and descriptive	: Output is more formal, structured and in numeric terms
Conclusions are drawn based on researchers judgment	: Conclusions are drawn on the basis of statistical tests

Types of case studies

a. ***Descriptive*:** This is the type of case study which is used to describe the phenomenon, its nature and to identify possible causative factors. It can be used in order to develop a working hypothesis for further research.

b. ***Exploratory*:** As the name suggest, exploratory case studies are used to identify and explore the natural phenomenon in order to understand its point of occurrence, frequency etc.

c. ***Prospective*:** Here a researcher uses a theory to predict the future occurrence of a natural phenomenon in the life of a social unit and evaluates to what extent does the theory hold true in reality and identify possible factors that interfere in theory.

Guba and Lincoln (1981) categorizes case studies into three types; factual, interpretative and evaluative.

Few of the frequently found applications of case study method in agricultural extension

a. To study entrepreneurial behavior

b. To explore successful farmers stories

c. To document ITKs.

Exercise

a. Identify three recently published and peer reviewed research articles which used the case study research method; and comment on research problem, type of social unit, number of social units selected, time coverage of study, nature of research outcome.

b. Enlist three examples of problem situations where case study is appropriate choice over statistical method.

c. Enlist three examples of problem situations where statistical method is appropriate choice over case study.

Reference

Yin, R. (2003). *Applications of Case Study Research* (2nd ed.). Thousand Oaks: Sage Publications.

PART III: TOOLS AND TECHNIQUES OF DATA COLLECTION

Chapter 16

Tools and Techniques of Data Collection: An Introduction

Scientific research is always based on observable and measurable facts. This applies to physical as well as humanitarian fields of science. In order to investigate social or psychological phenomenon, most of the times we need quantified information related to the phenomenon under investigation. In order to secure this data, we need one or more method/ means of collecting observable/ measurable data.

Data collection methods are systematic and standardized procedures for observation and recording of data pertinent to objective of research.

Types

i. Interview

ii. Mailed questionnaire

iii. Objective methods

iv. Projective methods (Tests and scales)

v. Content analysis

vi. Observation method

vii. Sociometric methods.

Chapter 17

Interview Schedule and Mailed Questionnaire

Interview method: It is a two way interactive (either face-to-face or through multimedia technology), method of data collection wherein the researchers elicit required response from the respondent by presenting stimulus in the form of series of questions pertinent to it.

Interview schedule: It is a data collection tool consisting of properly worded, sequentially arranged questions, tests relevant to objective of research.

Various forms of Personal interviews

a. ***Structured*:** Here the wording, content, and sequence of the questions/stimuli to be asked (used) are predetermined. It is more rigid and the interviewer is supposed to follow the interview schedule as such.

b. ***Unstructured*:** Here, the researcher has flexibility to collect the information by asking questions according to answers of respondent. There is no rigidity with respect to sequence and wording of the questions to be asked but should elicit required information.

c. ***Semi-structured*:** Here the researchers have flexibility to an extent for moderating the wording and sequence of questioning the respondent. But every question listed in the interview schedule must be asked to the interviewee. In extension research, most of the times we use semi-structured interview schedule.

Advantages

a. 'Difficult to understand' type of questions can be clarified

b. This technique can be used even with illiterates

c. There are very less chances of non-response and incomplete information
d. Response can be cross-checked by asking cross-question
e. Response are generally more accurate
f. Response can observe physical reaction pattern of respondent.

Limitations

a. In a given time period, with this method, information can be collected with relatively less number of respondents
b. More costly and time consuming
c. Requires skill on the part of the interviewer and researcher
d. Possibility of researcher bias in collection and recording of information
e. Not suitable for personal information and it reduces anonymity.

Mailed questionnaire

It is a method of data collection wherein self-administrable survey forms (questionnaires) –containing pertinent research questions –are mailed to respondent and respondent fills and sends back the filled questionnaire.

Advantages

a. Large coverage
b. Cost effective
c. Suitable for asking sensitive questions
d. Preserves anonymity
e. Suitable when respondents are dispersed on larger geographical area
f. Respondent can fill on his own convenience
g. Free from interviews bias
h. Does not require much skills for administrators.

Limitations

a. Cannot be used with illiterates
b. Low rate of returns
c. Difficult questions cannot be classified and so is misunderstood

d. High number of incomplete questionnaires

e. Delayed responses and follow-up is required

f. Chances of proxy response.

Interview schedule		Mailed questionnaire
Two-way interactive communication	:	Mostly one-way and direct
Response to stimulus is quick	:	Delayed response
Can be used with illiterates	:	Cannot be used with illiterates
Require minimal time	:	Delayed
Less number of respondents covered	:	Large number of respondents covered
Response recorded by interviewer	:	Recorder itself records response
Spontaneous response	:	Delayed response
Direct communication	:	Indirect communication

Steps involved in construction of Interview schedule/ questionnaire

a. Decide objective

b. Decide upon the variable/ psychological domain to be probed

c. Decide method of data collection (Interview schedule/ questionnaire)

d. Decide upon nature of data required- Based on this, decide type of questions to be included

c. Generate the questions that enquire domain under investigation

f. Prepare preliminary draft of Interview schedule/ questionnaire

g. Pretest of draft

h. Prepare final draft.

How to improve mailed questionnaire turnover

a. Send polite and frequent reminders

b. Ensure enclosure of necessary arrangements for dispatching of the completed questionnaire

c. Communicate the importance of their response to the respondents

d. Ensure that cover letter of the mailed questionnaire is having personal appeal

e. Attach informative incentives with questionnaire completion.

Checklist for developing and selecting questions

a. Questions should be clear and unambiguously stated
b. Avoid social desirability questions as much as possible
c. Question should be relevant to the objectives of research
d. Avoid judgmental questions whenever possible
e. Language and context of the question should be in line with knowledge and experience level of respondents
f. Avoid leading questions
g. Way the question is asked must elicit the response in a manner appropriate to nature of data required for statistical analysis
h. Use simple specific and decent language
i. Keep one idea only in each question
j. Use simple sentence in local vernacular language
k. Avoid repetition of questions
l. Add control questions.

Types of questions in Interview schedule/ Questionnaire

a. *Open ended:* Respondent have freedom to express their own response for given stimuli.
b. *Close ended:* Respondent is forced to choose one among limited number of alternatives provided.
c. *Scales:* Here respondent indicates his opinion attitude on a psychological continuum.

Open ended Questions	Close ended Questions
Gives true response of respondent	: Uniform data
Respondent can answer in his own terms	: Easy to code, categorize and tabulate; Easy to analyze
Reveals newer dimensions of response/ researchers may not be aware	: Conclusions can be drawn with wider generality
Increased involvement of respondent	: Keep the response concrete
Clears up misunderstanding of respondent	**Limitations**
In-depth information	: Non-exhaustive response
Limitations	: May not represent true response of respondent
Non-uniform data	: Superficial data
Difficult to code, categorize, order and tabulate	: Less involvement of the respondent
Difficult in statistical analysis	: Less chances of careless response
Judgment may be left to the researcher	: Less infrastructure

Exercise

a. Differentiate between tools and techniques.

b. Develop questionnaire for collecting the response from respondents with respect to at least five variables related to research problem already selected by you in previous sections (Except binary variables).

c. Select a thesis from library and write a comment on its questionnaire/schedule (500—700 words). Comment should include the variables measured, scales/tests used, strengths and limitations which you found in the selected questionnaire/schedule.

Reference

Kerlinger, F. (1973). *Foundations of behavioral research* (2d ed.). New York: Holt, Rinehart and Winston.

Chapter 18

Observation Method

It is the method of gathering the data by seeing, hearing and then recording the observation on the phenomenon/behavior of the respondent rather than the subjects' self-reporting of the response.

Kinds of Observation

Method of observation can be of one of four kinds (based on participation of the observer and the observation settings).

***Participant observation*:** In this type of observation, the observer becomes a part of the group/phenomenon which he/she is observing. The main advantage associated with this type of observation is that, one can get intimate knowledge about phenomenon under investigation under natural settings. But there are chances of observer getting involved emotionally and report biased observations. Additionally it demands lot of time and resources. Maintaining psychological distance is a challenge here.

***Non-participant observation*:** Here observations are taken by observer without becoming part of the phenomenon / group under investigation. Observer maintains physical and psychological distance from involvement with group under observation. Main advantage of this method over participant observation is that, one can collect observations more objectively due to non-involvement with the group under investigation. But this method is not as efficient as participant observation in collecting observations on intimate aspects of social life.

Non-controlled observation: Here observations are made under natural settings and phenomenon under observation occurs without being influenced by any manipulation from external force.

***Controlled observation*:** Here the observation takes place either in controlled environment or with definite pre-arranged plan involving experimental procedures in order to check biases due to wrong perception, incomplete information and effect of extraneous stimuli. Here awareness of the respondents about being part of experiment may lead to hawthorne effect and affect the external validity of the observations.

There are two types of controlled observation. First one is *control over phenomenon and second one is control over observation.*

***Control over phenomenon*:** In former case, the control is placed on the phenomenon so as to observe for the influence of specific factor / independent variable on the occurrence, intensity, frequency and extensity of the phenomenon under observation.

***Control over observation*:** In later case the control is placed on the observer and observation method (for example using of check list, team observation, mechanical appliances etc.) so as to avoid biases and human errors in observation.

Exercise

Review the recently published research articles and discuss in detail at least five techniques used by various researchers for controlling observation bias.

Chapter 19

Content Analysis

Communication material is the joint product of knowledge, skill and attitude base of the communicator. Further, most of the communication done on behalf of certain organization also reflects the objective, work culture, strategy and contribution of the organization. Hence analysis of communication material is of greater relevance in exploring the trends, vision, behavior and future course of action. In this chapter, we are going to discuss the content analysis; it's meaning and practical application.

Content analysis is a method of analyzing the communication content and treatment in a systematic, objective and quantitative manner to measure variable aspects of communication.

Practical utility of content analysis in extension education

a. To evolve and evaluate communication standards.

b. To discover technologies of propaganda.

c. To analyze the subject coverage in communication.

d. To evaluate readability of communication material.

e. To discover communication intensions.

f. To measure attitude, interest and values.

Steps involved in content analysis

Step		Example
Define the universe of communication to be analyzed: It is based on objectives of the study, availability of material.	:	Community radio programme on agriculture
Select the sample	:	60 radio programmes were selected randomly
Categorization	:	Selected communications were categorized into four categories; Radio talks, Phone-in programmes, agricultural folk songs and dramas
Select unit of analysis	:	Here unit of analysis is the theme of the programme
Coding and analysis	:	Coding each of the identified categories and developing of descriptive statistical data (Example: Frequency, Percentage etc.)

a. **Categorization:** Based on object of study, categorize the sample context into meaningful categories. Categories should be exhaustive and mutually exclusive. Example: Radio programmes can be classified into agricultural, horticultural, floricultural aspects etc.

b. **Decide upon Unit of analysis:** It refers to measure in terms of which analysis can be done. It can be a word, a theme, an item or a character.

   Word: Meaningful and significant word relevant to objective. Example: Yield

   Theme: Theme is an idea expressed in 2 or few words or in 1-2 sentences. Example: Genetically modified crops

   Item: Item refers whole communication response product of a subject towards a given stimuli. For example: Feedback letter written by the farmer to the broadcaster.

   Character: Here, words rather than content, the unit of analysis is a person/character. Example: Achilles in Roman literature; scientist giving a radio programme is a unit of analysis.

c. **Coding and analysis:** Coding is a process of assigning symbols to units of analysis according to attribute and then categorizing them. In analysis, we summarize the communication content with respect to the communication characteristic of the units of analysis.

Systems of enumeration in analysis of communication content

a. **Appearance:** It is a measure of importance of content units. It analyzes in what proportion of selected sample communication, the content unit appears.

b. **Frequency:** Number of times a particular unit of analysis appears in communication content. Example: No. of times Genetically modified (GM) crops appear in radio programme.

c. **Coverage:** Location of unit of content in communication material and amount of space or time covered by it. Example: No. of hours of lectures on GM crops.

d. **Information content:** Measure of relevancy of content.

Advantage

a. Applicable to almost every type of communication.

b. Free from response bias and willingness of respondent.

c. Can be carried out at researcher's convenience.

d. Can be easily checked for accuracy of recorders information.

e. Appropriate for open ended investigations.

Chapter 20

Sociometry

Existing positive as well as negative interactions among the members of the society determine the communication structure, leadership and technology diffusion pattern within the given social system. Sociometry is a method for gathering and analyzing data on communication choice and interaction pattern of individuals, within a group in order to identify communication structure and pattern.

Sociometric methods refer to a large class of methods that assess the positive and negative links between persons within a group. The basic principle of the sociometric method is that every group member has the capacity to evaluate every other member on one or more criteria in a round robin design (Rubin, Bukowski and Laursen 2009).

Data generated in Sociometric analysis can be analyzed in different ways. Of these, most relevant ones in extension research are:

a. Sociometric Matrices

b. Sociometric Indices

c. Sociograms (Directed Graphs)

Sociometric matrices: It is a numeric representation of communication choices of individuals within a group/community in a matrix form. This represents the whole set of data on all the individuals' choice pattern in tabular format. Note that the format of table varies depending on whether there is a limit on number of choices for each individual or not. Below given is the example of the Sociometric matrix without any limit on number of choices. In this hypothetical example total number of members are seven, so for each individual maximum six choices are possible. This example is of limitless choices, because each of the members within the group provides his/her response about his/her communication with rest of every member within the group. Alternatively, we can also ask every individual to choose only two members with whom he/she interacts most frequently. Under this condition it will be the limited choice model of Sociometry.

In below given table, '1' indicates that individual communicates frequently (or prefers to communicate) and '0' indicates that the given individual does not communicate frequently (or do not prefer to communicate) with the person in corresponding column.

	A	B	C	D	E	F
A	—	1	0	1	1	1
B	1	—	0	0	0	0
C	0	0	—	0	0	0
D	1	0	0	—	0	1
E	1	1	0	1	—	1
F	1	0	0	0	0	1
G	1	1	1	0	1	—
Total Choices	5	3	1	2	2	4

We can use Sociometric matrix to identify choice star, secondary choice star and isolated individuals.

Choice star: The person who obtains highest number of communication preference choices from the rest of the members within the group is the primary choice star. Person 'A' in the above example obtained highest number of choices; Hence A is the choice star in this case.

Secondary choice star: The person who obtains second largest number of communication preference choices from the rest of the members within the group is the secondary choice star. Person 'F' in the above example obtained second highest number of choices; Hence F is the secondary choice star in this case.

Isolated individual: Person who obtains least number of communication choices from the rest of the members within the group. Person 'C' in the above example obtained least number of choices; Hence C is the isolated individual in this case.

Application of Sociometry

1. Identification of leader.
2. Categorization of population based on communication pattern.
3. To identify communication structure.
4. Identify isolated individuals in group.
5. To identify clique.

Sociometric Indices: These are derived from ratio of various types of choice patterns to the total number of corresponding possible choices. There are two major indices used in extension research –Choice status and Group cohesiveness.

Choice status: It is the ratio of number of times an individual is chosen to the total number of possible choices.

Choice status $= \frac{\Sigma C_j}{n-1}$

Where, ΣC_j = Number of times the jth individual is chosen

n = Total number of individuals within the given group

Note that, value of choice status index is always smaller than or equal to unity. Larger the value of choice status, higher is the communication preference of the group for the given individual. While using the Sociometry for identification of group leader, generally the person with highest choice status index value is considered for leadership.

Group cohesiveness: It is the index which measures extent of mutual communication and communicational cohesiveness among the members within the group. It is calculated by using the below given formula.

Case I: When there is no limit on number of choices to be made

$$Group\ cohesiveness = \frac{\Sigma(i\langle-\rangle j)}{\frac{n(n-1)}{2}},$$

Where, $\Sigma(i <-> j)$ refers to the total number of mutual choices made by the individuals in the group.

Case II: When there is no limit on number of choices to be made

$$Group\ cohesiveness = \frac{\Sigma(\mathbf{i} <-> j)}{\frac{\mathbf{dn}}{2}},$$

Where *d* is the maximum number of choices allowed.

Sociograms: Sociograms are the directed graphs which diagrammatically represent the communication choice pattern of the individuals within the given group. These are very useful in identification of Cliques within group. These diagrams become more complex and difficult to understand and interpret manually as the number o individuals within the group increases. Sociograms are also useful in analysis of communication structures. There are two types of communication choices within the Sociograms; One way choices, indicated by

"→" and mutual choices indicated by "↔". Figure is an example of Sociograms based on the communication choice pattern among the members in a group of size six (i.e. A, B, C, D, E and F).

From the figure it is quite clear that each of these four members (A, B, D and E) mutually interact with each other, hence there exists a clique of four members in the group. Further it is also obvious that individual F is an isolated individual because he/she has only few communication preferences.

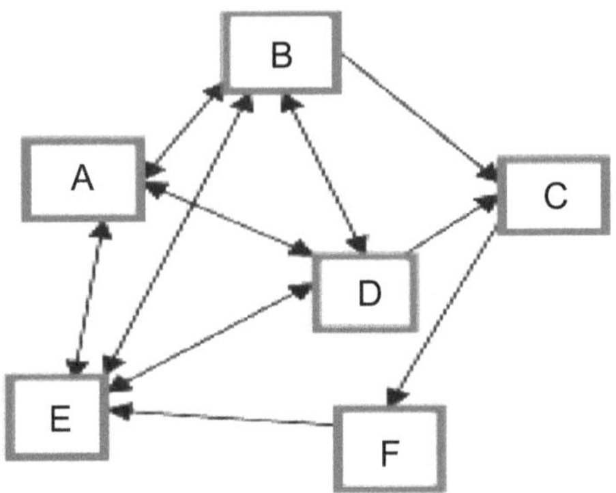

Applications of Sociometry

a. Identification of primary and secondary choice star.

b. Identification of group leader.

c. Categorization of the community based on communication patterns.

d. Illustration and analysis of complex communication structures.

e. To identify of Clique within the group.

f. For identification of isolated individuals.

Exercise

1. Use sociogram to design communication and interaction pattern in your course batch. Identify if there exists any clique.
2. Discuss various other Sociometric methods (other than those mentioned in this manual) used in communication and interaction studies (500-700 words)

References

Rubin, K., Bukowski, W. M. and Laursen, B. (2009). *Handbook of peer Interactions, Relationships, and Groups*. New York: Guilford Press.

Kerlinger, F. (1973). *Foundations of Behavioral Research* (2d ed.). New York: Holt, Rinehart and Winston.

Chapter 21

Projective Techniques

It is not possible that every respondent is willing to and capable of giving honest and true response. Sometimes, it is practically not possible to make the respondent reveal about personal information, but true research requires true response. When direct stimuli fail to elicit true response, then we must use indirect and extractive methods to elicit the required response indirectly from the respondents, which further can be interpreted in the relevant format. This indirect and ambiguous stimulus based methods are known as projective methods. *Projective methods are technique of data collection on the person, value, attitude, emotions and motives etc. by using a highly unstructured and ambiguous stimulus* (Wilkening, Wilkening & Wilkening, 1973).

Principle: Principle of projective technique is that more unstructured and ambiguous the stimulus is, more the person will project his emotions, motives, values and attitude.

Characteristics of projective techniques

a. These methods utilize ambiguous and highly unstructured stimuli.

b. Basic interferences are drawn on the basis of judgmental interference.

c. Almost anything can be used as stimuli.

d. Since each judge can judge the response based on his own experience, the objectivity in the obtained information in relatively lower.

e. Requires skill and experience on the part of response interpreter.

f. Relation and validity hence doubtful.

Classification of projective technique (Based on types of response; a/c to Lindzey)

a. Association

b. Construction

c. Completion

d. Choice/ ordinary

e. Expression

a. ***Association technique*:** In this, the person is presented with stimulus and asked to express what first comes to mind when he/ she receives the stimuli. Based on respondents association of stimuli that is different thoughts, words that comes spontaneous in their mind; the interference is drawn. Example of Association technique is the "Word association technique".

b. ***Construction technique*:** Product of the subject will be the aspect interest in this case . Here the respondent is exposed to stimuli (mostly picture/ image) and he is supposed to construct some product (mostly a story) based on the stimuli. Thus constructed output is interpreted by the researcher. Example: TAT –used by McClelland to measure achievement motivation.

c. ***Completion technique*:** Here the subject is supplied with stimuli that are incomplete in some form. The subject is suppressed to complete the stimuli in the way he wishes to. Example: Sentence completion technique

d. *Expression technique*: It is also similar to construction technique but the focus in it is not on product but on manner/ style of framing the product/ expressing it. Example: Playing, drawing, painting etc.

e. ***Choice/ ordering technique*:** Here subject chooses the item amongst several alternatives, each of the item indicates different degree of achievement, motivation or attitude towards the objectives etc.

References

Lindzey, G. (1959). On the classification of projective techniques. *Psychological Bulletin*, 56(2).

Wilkening, H. and Wilkening, G. (1973). *The Psychology Almanac: A handbook for Students*. Monterey, Calif.: Brooks/Cole Pub.

PART IV: DATA PROCESSING AND REPORT WRITING

Chapter 22

Coding of Data

Data collected by using different data collection methods remain meaningless until it is screened, cleaned, coded, analyzed and summarized. Data cleaning and screening mostly come under the purview of statistical logics and techniques hence not dealt with in this practical guide. Since these aspects will be dealt by the statistical experts from the allied statistical wings. Presuming that students are aware about data cleaning and screening, we will proceed further with data coding.

Coding is the process of categorizing raw qualitative data into meaningful analytical units and assigning codes or labels.

Purpose of data coding

a. To aid in summarizing the qualitative data.

b. To ease in the access to and understanding about huge amount of collected data.

c. To aid in qualitative data analysis.

Types of data coding

Corbin (1990) classified coding into two types; Open and Axial coding.

***Open Coding*:** Glance through the transcript/information and mark the sections of the text that represents/ depicts the behavior/ phenomena under investigation. For example, you circle words or phrases describing the behavior of the course instructor.

***Axial Coding*:** When you have large number of codes, then it is necessary to sort the identified codes into different groups. This is called *axial coding*. Two frequently used types of axial coding are Hierarchical and Non-hierarchical coding.

***Hierarchical Coding*:** Here you find different codes that can be grouped together, but they can be further sub-grouped into hierarchy based on certain logical criteria. For example: When we are coding type of friendship among adults, we can code the responses in following format.

Example: Types of relation

- Informal friendship
  - Close friend
  - Sporting
  - Pen friend
- Formal relationship
  - Superior-subordinate
  - Co-worker
  - Subordinate-superior

Non-Hierarchical: This is a type of coding in which the responses of the subject are not hierarchically arranged but they are simply grouped / categorized.

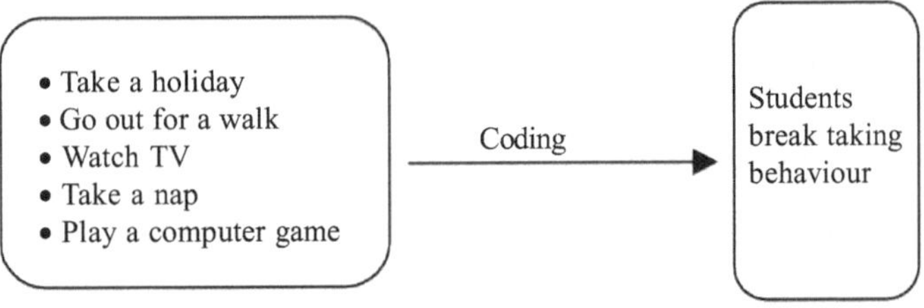

Few examples of most frequently coded qualitative information

***Behaviour*:** Example: Students break taking behaviour.

Events: Important days in life.

***Activities*:** Students' time spending behaviour.

***Strategies*:** Students' studying strategies.

***States*:** Frustration, Disappointment.

***Meanings*:** Perception regarding climate change.

***Relationships*:** Close, Distant, formal *etc.*

***Socio-personal characters*:** Gender, Social contact *etc.*

***Settings*:** Homely, work place, Casual meeting place.

***Consequences*:** Good, Bad, desirable, undesirable *etc.*

Chapter 23

Testing of Hypothesis

Hypothesis testing is the statistical procedure used in order to accept or reject statistical null hypotheses. One of the major purpose for which we formulate the hypothesis is, to test the statistical significance of the relation, difference or variation etc. among given sample/s at a given point of time or at different time points.

Hypothesis testing or significance testing is a method for testing a claim or hypothesis about a parameter in a population, using data measured in a sample. In this method, we test some hypothesis by determining the likelihood that a sample statistic could have been selected, if the hypothesis regarding the population parameter were true (Privitera, 2012). *One should be very clear that we do neither test the significance of null hypothesis nor the significance of alternate hypothesis, but we subject null hypothesis to the process of statistical testing in order to test the significance of relationship between two or more variables. We do not test the variables per se but the relationship that exists between them.*

Steps in Testing of hypothesis

1. Review the literature, synthesize from culture, experience and analogy and formulate the scientific problem.
2. Restate the scientific problem in the form of testable null hypothesis.
3. State alternate hypothesis.
4. Identify sampling frame and select the sample.
5. Selection of test statistic (f, t, z etc.).
6. Decide upon level of confidence (5%, 1% etc.).
7. Collect and analyze data; calculate test statistics.
8. Make decision and draw conclusion.

Level of Significance and confidence level

Note that significance level should not be confused with confidence level. Both give almost the opposite meaning. Significance level is the probability of rejecting null hypothesis when it is true, whereas ***confidence level*** is probability of accepting null hypothesis when it is true and is calculated as 1 - α, where α is the significance level. Both the values –confidence level and level of significance –are generally expressed in terms of percentage. Generally 1%, 5% levels of significance are used for testing of hypotheses in social sciences; very rarely 10% and 15% level of significance are used in social sciences. When we wish to accept the alternate hypothesis only with least possibility of any deviations from it, then we prefer smaller level of significance, because we want to give minimum margin to the deviation from our experimental results. In medical sciences, Alpha values are used far smaller than those used in Social sciences.

Remember that, sum of significance level and confidence level is always unity.

$(\alpha) + (1\text{-}\alpha) = 1$

Significance level + confidence level = 1

Major basis for statistical testing of any hypothesis is the phenomenon of probability. The interpretations drawn from any statistical testing is mainly based on the whether a null hypothesis is accepted or rejected. Subsequently, whether a null hypothesis is accepted or rejected depends on the level of significance and the confidence we wish to have in our research findings (i.e. confidence level).

Interpretation of the results: Note that we test the deduced statistical implications of null hypothesis and not the alternate hypothesis. A statistical hypothesis is an assumption about a population parameter. This assumption may or may not be true. We will accept only the true null hypothesis and reject the false ones. Conclusion regarding acceptance or rejection of hypothesis depends on value of test statistics and level of significance.

Test statistics may be t, F, z etc. but the interpretation of these test statistics depend on the value of calculated test statistics and the critical value of test statistics (based on the value of level of significance). Critical test statistic value for the given sample size (i.e. degrees of freedom) and the given level of significance can be located by using test statistic distribution table (t table, *z* table etc.). Once the critical value of test statistics is located, one can use the thumb rules mentioned below for making decision on hypothesis testing.

Traditional method

i. If calculated value of test statistics is larger than the critical value of test statistics then reject the null hypothesis.

ii. If calculated value of test statistics is smaller than the critical value of test statistics then reject the null hypothesis.

Modern method

The rationale behind this method is that, we will reject the null hypothesis if theprobability of committing type I error (*p value*) is larger than the maximum acceptable limit of committing type I error (i.e. level of significance, α). The *p value* can be generated for corresponding calculated test statistics either manually or by using data analysis softwares like SPSS, SAS etc. We will not discuss further on the process of manual calculation of *p value,* since the decision done by using both the traditional and modern method are same and most of the softwares provide the p value for the corresponding test statistics values. Decision rules for accepting or rejecting null hypothesis are given below.

a. If $p > \alpha$, then accept the null hypothesis

b. If p d" α, then reject the null hypothesis

Another issue associated with the decision making based on level of significance is the issue of nature of hypothesis. If the hypothesis is non directional then we will use two tailed test, splitting the level of significance equally on both the halves of the test statistics distribution.

In this case, critical value is a cutoff value that defines the boundaries beyond which less than 5% of sample means can be obtained if the null hypothesis is true. Sample means obtained beyond a critical value will result in a decision to reject the null hypothesis. The regions beyond the critical values are called the rejection regions. If the value of the test statistic falls in these regions, then the decision is to reject the null hypothesis; otherwise, we retain the null hypothesis.

Figure 1 represents the test statistics distribution for difference between two mean. Suppose we set the maximum acceptable limit of type I error to 0.05, then level of significance (α) in this case is 5%. Since our hypothesis is non-directional hypothesis, we have 2.5% of the area in the upper and lower tails as an area of rejection. Rest of the 95% of the area is area of acceptance. The upper and lower limit of region of acceptance is called as critical test statistics value. Thus, we need to find the critical values between which lie 95% of the statistical values on either side of the mean. Any observed value that exceeds those values in either the positive or negative direction leads to rejection of the null hypothesis.

Rejection regions: Observed test statistics values in these regions lead to rejection of the null hypothesis because the error probability in these cases is lower than our maximum set limit for type I error (i.e. α).

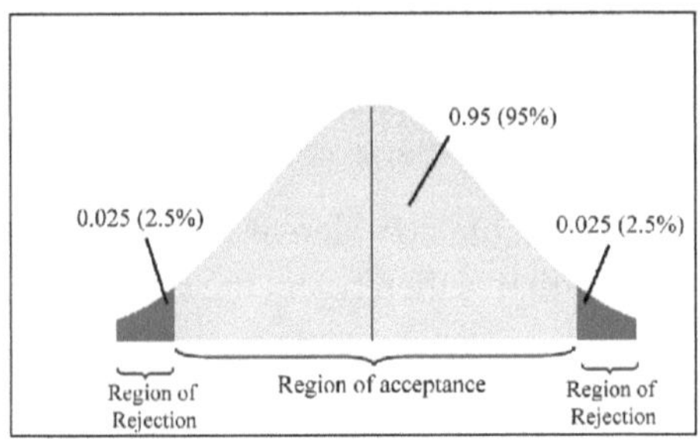

Fig.1: Left tailed test (H_1: $\mu < \mu_0$)

The previously stated situation was the case of non-directional hypothesis, where we were interested only in testing the statistical significance of difference between two means/ proportions /variance, without specifying which one is supposed to be larger or smaller. Now, we will look into the case of directional hypothesis where we specify that, which one of the two samples being compared have smaller or larger mean/ proportion. This is also applicable to single sample test, where we compare the observed sample mean / proportion with the some presumed value of mean / proportion.

Note that, there are two possibilities in statistical testing directional hypothesis. First one is left-tailed test (where alternate hypothesis is in the form of $\mu< \mu_0$) where the rejection region lies entirely on the left tail of the sampling distribution (Figure 2). Second possibility is right tailed test (where alternate hypothesis is in the form of $\mu< \mu_0$). In this case the rejection region lies entirely on the right side of the test statistics distribution (Figure 3).

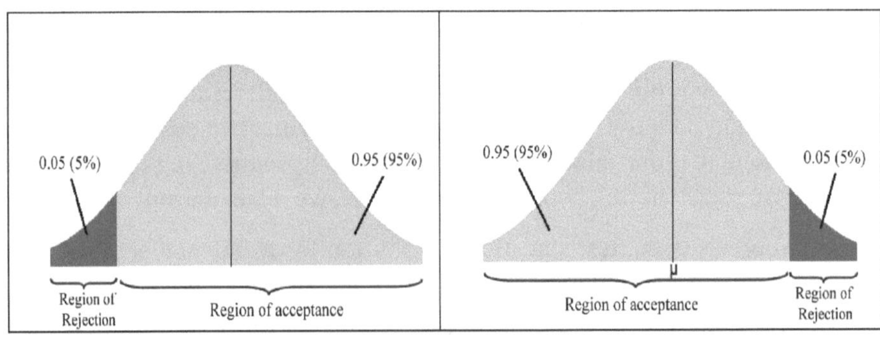

Fig. 2: Left tailed test (H_1: $\mu < \mu_0$)

Fig. 3: Left tailed test (H_1: $\mu > \mu_0$)

Decision errors in hypothesis testing: Whenever we test a hypothesis and calculate the test statistic for given set of observations, and then four decision scenarios are possible.

a. Null hypothesis is true in reality and based on statistical testing, we decided to accept null hypothesis.

b. Null hypothesis is true in reality whereas we decided to reject the null hypothesis based on statistical testing.

c. Null hypothesis is false in reality and we decided to accept the null hypothesis based on statistical testing.

d. Null hypothesis is false in reality and we decided to reject the null hypothesis based on statistical testing.

Note that decision A and D are correct ones whereas the decisions B and C are erroneous decisions because they do not represent reality. Based on this pattern of decision making, there are two types of errors in decision making regarding testing of hypothesis.

i. **Type-I Error**: It is the error of rejecting of null hypothesis when it is true. Probability of committing type-I error is denoted as a. The largest probability of committing a Type I error that we will allow and still decide to reject the null hypothesis is referred to as Significance level (Privitera, 2015). Type I error generally leads to conclusion that significant relationship exists between two variables, when in fact it doesn't. *Significance level* directly controls for the probability of committing type I error by setting the maximum acceptable limit on the probability of committing the type I error.

   Example: A test concludes that there exists significant relationship between achievement motivation and economic status, when in reality no such relation exists between achievement motivation and economic status. This error of statistically establishing a significant relationship between two variables, when it doesn't exist in reality is a kind of Type-I error.

ii. **Type-II Error**: The error of accepting of false null hypothesis is referred to as type-II error or b error. Probability of accepting a false null hypothesis is denoted as â.

   Example: A researcher carries out an experiment to determine the relationship between amount of fertilizer applied and the paddy crop yield. Suppose in reality there exists a positive correlationship between the amount of fertilizer applied and the paddy crop yields, whereas the test proves that there exists no significant correlationship between the amount of fertilizer applied and the paddy crop yields, then we are falsely accepting the null hypothesis (or hypothesis of no significant correlation). This type of error is referred to as type –II error.

		Statistical Test Decision	
Reality	H-$_0$ is True	*Accept H_0* Correct decision; (1-α) Probability of this decision is called as *Confidence level*	*Reject H_0* *Type I error (α)* Probability of this decision is called as *Level of significance*
	H-$_0$ is False	*Accept H_0* Type II error (β)	*Accept H_0* Correct Decision; (1-β). Probability of this decision is called as *Power* of test

Power of a test

The power of a statistical test is the likelihood of rejecting the null hypothesis when is false. This is denoted by 1-β. It is called power because it is the decision we aim for. Remember that we are only testing the null hypothesis because we think it is wrong. Deciding to reject a false null hypothesis, then, is the power, in as much as we learn the most about populations when we accurately reject false notions of truth. This decision is the most published result in behavioral research (Privitera, 2015)

Importance of Power of test

Although one can carry out testing of hypothesis without calculating the power of test, estimation of power of a test in advance will help to ensure that the selected sample is large enough for the purpose of the test. If the sufficiency of size of sample is not ensured then the test may be inconclusive, leading to wastage of resources and indecisive conclusions. Occasionally the power of a test is calculated after performing the test, but this is recommended only to determine adequacy of sample size in a follow-up study (suppose a test failed to detect an effect, it was obviously underpowered – nothing new can be learned by calculating the power at this stage). Power of a test is useful in estimating the required sample size for a given research.

Factors affecting power of a test

i. Sample size (+)

ii. Sample heterogeneity (-)

iii. Effect size to be detected (+)

iv. Significance level (alpha) of the test (+).

Note that, positive sign in parenthesis indicates a directly proportional relationship between power of a test and the corresponding factor, whereas negative sign indicates inversely proportional relationship.

Exercise

1. Briefly answer the following questions

a. What type of error is associated with decisions to retain the null?

b. What type of error do we directly control?

c. What type of error is associated with decisions to reject the null?

d. State two correct decisions a researcher can make in testing of hypothesis?

2. Indicate whether following statements are true or false. Justify your answer in 3-4 sentences.

a. The probability of *not* committing a Type II error is called the **Power** of the test.

b. Keeping the value of calculated test statistics constant, acceptance or rejection of a given null hypothesis depends on the required confidence level.

c. Aim of hypothesis testing is to make conclusion regarding acceptance or rejection of alternate hypothesis.

d. One tailed tests are used for testing non-directional hypotheses.

3. Complete the sentence

a. In a study aimed at assessing the effect of new instructional method on the speed of learning, researcher found that new method of instruction have statistically significant impact on speed of learning at 1% level of significance. This means that confidence level is ________%.

b. For an upper-tail critical hypothesis test, the rejection region lies on the __________ side of test statistics distribution.

c. For a lower-tail critical hypothesis test, the rejection region lies on the __________ side of test statistics distribution.

— —

— —

— —

— —

— —

Chapter 24

Thesis Writing

Based on the nature and purpose of the content, there are numerous classes of scientific communications. Theses represent the class of research communications that are of great importance to the students. Thesis is part and parcel of the master's and doctoral programme in almost every academic institution. In some institutions, even the bachelor students are expected to submit their brief report on project.

Thesis is the detailed written communication of the problem statement, abstract of the research process and results, the literature which helped the researcher in designing and supporting his study, the research methods which researcher used as a means of manipulation, control, observation, measurement and data analysis and the results of the research work along with its discussion and the bibliography of the reference material used for designing of the study and for supporting of the results of study.

Format of the thesis varies from university to university. But review of the thesis format used by various universities shows existence of certain common components in every thesis. These common components of the thesis are:

1. Title Page
2. Acknowledgement
3. Certificates -of authentication and examination
4. Table of contents
5. Abstract - of the research process and results
6. Introduction - to the study and the problem statement
7. Review of Literature
8. Research Methodology/ Methods and Materials

9. Results and Discussion
10. Summary and Conclusion
11. Bibliography
12. Vitae.

Actual thesis may show deviations from the above listed components. The deviations will be minor. Sometimes even though the wording (naming) pattern of the component differs but the name of the component gives same meaning. Below given is the description chart about the each of the component.

Component	Description
Title	• It should be clear, concise and specific • It should denote the central ideas of your thesis • It must imply what your argument will reveal about your topic • Abbreviations should be avoided in titles unless a scientific community is better aware by its acronym • Ideally, length should not exceed twenty (20) words
Acknowledgement	• Genuine technical, financial and educational contribution of the academic institutions, technical experts, funding agencies, facility providers and research aide should be acknowledged. • Acknowledgement should not be confused with the 'dedication'
Certificates	• Certificate of originality of the work (from author) • Certificate of research guide • Certification by advisory committee • Certificate of competent authority/head of the institution
Abstract	• Length should be limited to 250-500 words • It should summarise the research process and the findings of the research • It should cover entire range of important findings and should be conclusive • More emphasis should be on findings and conclusion rather than on research process • The abstract should also be written in the third person • Generally, the first sentence of an abstract describes the entire study; subsequent sentences expand on that description.
Table of contents	• Enlist the entire list of headings and subheadings along with the with their respective page numbers The table of contents may be followed by List of tables, List of figures, List of illustrations, List of symbols and list of abbreviations wherever necessary
Introduction	It should include brief details about

(Contd.)

Component	Description
	• Research problem • Problem statement • Statement of purpose • Rationale and significance • Overview of methodology • Expected outcomes • Assumptions of the study • Limitations of study
Review of literature	• Opening remarks should state the content and organization of the literature reviewed as well as the strategy used in the literature search • It should analyse and synthesise the relevant studies across discipline and compare & contrast different research approaches, outcomes, interpretations and conclusions • Review the primary and secondary source that are recently published empirical studies in scholarly journals and publications • Use archival records if the nature of research demands for it; for example case studies • Content should be logically organized by theme or subtopic, from broad to narrow Provides section summaries • Describe the Conceptual framework at the end of review of literature
Methodology	• Introduction to the research problem • Rationale of research approach • Context of the research • Operationalization of variables • Define population and entail the sampling procedure • Data collection methods • Data analysis methods • Limitations and dclimitations
Results and discussion	• Findings should be presented in clear narrative form using plentiful verbatim quotes, and detailed description • Findings should be supported by the previous studies • Deviants findings should be justified with alternative explanations • Findings should be logically arranged in the form the research proceeded • Discussion part should provide an in-depth analysis of the findings
Summary and conclusion	• This chapter reflects upon the findings of the study, interrelates the findings if necessary and provides meaning to the entire set of findings. • Draw conclusions on research problem based on the findings of the study.

(Contd.)

Component	Description
	• Set the stage for the further study by describing what was outcome of study and what further needs to be done for the advancement of science
Bibliography	• Enlists the citation for the reviewed literature, which appears all along the entire content of the thesis. • Citation should be written in the format prescribed by the university/institution. • Citation should appear in alphabetical order.
Vitae	• Briefs the authors information, educational background, area of expertise, contact information etc.

Exercise

a. Select a recently published thesis from library and evaluate the thesis content against the checklist table provided in this chapter.

b. Discuss in detail how thesis writing differs from technical report writing.

Reference

Bloomberg, L. and Volpe, M. (2012). *Completing your qualitative dissertation: A road map from beginning to end* (2nd ed.). Thousand Oaks, Calif.: SAGE Publications.

Zeitfracht Medien GmbH
Ferdinand-Jühlke-Straße 7
99095 Erfurt, Deutschland
produktsicherheit@kolibri360.de